THE ENERGY CRISIS

∾

Paul W. McCracken, *Moderator*

∾

Clifford P. Hansen
Morris K. Udall
Charles E. Spahr
Mike McCormack

Jennings Randolph
Mark O. Hatfield
Dixy Lee Ray
Philip H. Trezise

J. William Fulbright
John N. Nassikas
George W. Ball
Charles J. DiBona

An AEI Round Table held on 25, 26, and 27 September 1973
at the
American Enterprise Institute for Public Policy Research
Washington, D.C.

THIS PAMPHLET CONTAINS THE PROCEEDINGS OF
ONE OF A SERIES OF AEI ROUND TABLE DISCUSSIONS.
THE ROUND TABLE OFFERS A MEDIUM FOR
INFORMAL EXCHANGES OF IDEAS ON CURRENT POLICY PROBLEMS
OF NATIONAL AND INTERNATIONAL IMPORT.
AS PART OF AEI'S PROGRAM OF PROVIDING OPPORTUNITIES
FOR THE PRESENTATION OF COMPETING VIEWS,
IT SERVES TO ENHANCE THE PROSPECT
THAT DECISIONS WITHIN OUR DEMOCRACY WILL BE BASED
ON A MORE INFORMED PUBLIC OPINION.
AEI ROUND TABLES ARE ALSO AVAILABLE ON
AUDIO AND COLOR VIDEO CASSETTES.

ISBN 0-8447-2047-X
LIBRARY OF CONGRESS CATALOG CARD NO. L.C. 74-78404

PRINTED IN UNITED STATES OF AMERICA

PART I

BASIC ISSUES

Paul W. McCracken, *Moderator*

Clifford P. Hansen
Morris K. Udall
Charles E. Spahr
Mike McCormack

PAUL W. McCRACKEN, Edmund Ezra Day university professor of business administration, University of Michigan, and Round Table moderator: This evening we will deal with the basic issues of the energy problem.

I suppose if we had told the average citizen two or three years ago that we were going to have a discussion of energy, he might have wondered whether we were going to talk about old age or the desirability of vitamin pills. But over the last couple of years, we've had some education on what this problem of energy really is. At least all of us now have a nodding acquaintance with such things as BTUs, gasoline shortages, millions of barrels per day, lack of heating oil and the like.

The first question to which the panel ought to turn its attention is just how this situation suddenly erupted and what caused it. Senator Hansen, how would you answer that?

CLIFFORD HANSEN, United States Senate (Republican, Wyoming): Well, Dr. McCracken, I wouldn't say that the subject has suddenly erupted. If we look at the record, we will find that people in government and industry, and other experts as well, have been forecasting an energy problem for some time now on the basis of good, hard evidence. Declining reserves in this country, a falling-off in wells drilled, increases in demand, and other new factors gave those looking closely at the subject good grounds to believe that we were approaching an energy crisis.

3

But it isn't until somebody is unable to buy gasoline or unable to buy fuel oil that we really think there is anything like an energy crisis.

I am reminded of a story about an old farmer who said that he didn't know he was out of hay until he threw the last bale out of the barn. I suspect that, for the average American, all the talk, all the concern that had been expressed by others, and all the grim predictions just fell on deaf ears. It was only when people were unable to get either fuel oil or gasoline, or to catch a convenient airline flight because there was no jet fuel, that they were aware of a crisis.

MORRIS K. UDALL, United States House of Representatives (Democrat, Arizona): I would say that this has been building for about three decades—since World War II. The American people have been on a binge, a joy ride, and it is all over now. We will not have just a little crisis this winter and then find that everything is lovely again. We will be living with this thing for the next couple of decades, until we find some long-range solutions.

It is going to change the way Americans live, how they get to work, and the way they do things. It has been coming for a long time because of some exponential factors. It's impossible to double and then redouble demand and redouble your population the way we have and not suddenly wake up someday and find that the good old Exxon "tiger" is out of gas, or that the living room is going to be cold this winter.

I want to emphasize that this crisis is real, it is genuine. Americans are looking around for villains today. The environmentalists like to say, "The oil companies caused this shortage; they just made it up to get higher prices." And the oil companies like to say, "The environmentalists brought it on." Both of these views are false. The thing is very real.

We've all heard the old story about good news and bad news. Well, there will be two kinds of news in the next few years, bad news and terrible news. There just isn't going to be any good news. It is all bad news. We are in trouble.

CHARLES E. SPAHR, Standard Oil Company of Ohio: I think we might admit that those of us who have been doing the estimating were off about a year or so in the timing of our predictions. We simply couldn't foresee the acceleration in demand and usage caused by a number of factors that simply didn't exist a few years ago.

DR. McCRACKEN: In other words, we are consuming in 1973 the amount of oil we'd been predicted to consume in 1974 or 1975?

MR. SPAHR: That is about it.

MIKE McCORMACK, United States House of Representatives (Democrat, Washington): I would agree. This is certainly not a short-term phenomenon; it is a long-term phenomenon and it will be with us for the rest of this century. As Mo Udall said, it is a situation where we are dealing with exponential curves. The combination of a tough winter and the incremental impact of environmental laws have forced demand to leap across the supply line a year or two sooner than we expected.

DR. McCRACKEN: We hear a great deal about the fact that refinery capacity is now quite short. As I recall it, American refinery capacity has been rising rather slowly, at least relative to Europe and elsewhere. How come?

MR. SPAHR: Well, Dr. McCracken, refining capacity had been rising yearly until a few years ago, but it never got very far ahead of demand. I think it is fairly accurate to say that capacity kept about 5 or 6 percent ahead of demand each year. But since demand has been increasing in recent years at about 4 percent annually, we've really been only about a year and a half ahead of requirements.

Now, as long as there were no unusual impediments in the way of increasing capacity, as long as that extra capacity could be justified economically, this apparently small margin was enough to take care of things. Then a number of things started to happen. As you know, the new National Environmental Protection Act was enacted in 1970. The interpretation of that act led to a number of decisions which caused some shifts in usage from one kind of fuel to another. Demand accelerated but additions to refinery capacity tended not to. Our refiners were wondering what the final requirements for product specifications were going to be; they were concerned about how much money they would have to spend revamping their old plant in order to make the kind of fuel that would be required in the new era, and they were concerned about money for adding new plant capacity. So, they had to be conservative.

They were also beginning to worry about supply. The United States was getting close to the time when our own internal oil supplies would be nowhere near great enough to satisfy refining capacity. And we still had oil import controls. Thus a refiner looking to foreign oil sources for justification for a new plant, but unable to get a new quota because he hadn't been importing the oil needed to get a new quota simply couldn't do anything. He was dealing with other people's money and couldn't justify taking extraordinary risks.

So a combination of things, only a few of which I have mentioned, slowed the normal progress and helped us into the kind of fix we are in today.

DR. McCRACKEN: It's been my impression that one aspect of the energy problem has to do with slippage in the expected scheduling of nuclear power facilities in the electric power area. Is that true? Congressman McCormack, I believe you are on the Joint Atomic Energy Committee.

CONGRESSMAN McCORMACK: Yes, this is true—to a limited extent. I believe we have a dozen to fifteen reactors fewer on the line today than we had expected to have. But we have been getting new reactors on the line and more will be coming. The slippage has been one to two years.

Certainly, this is a factor, but it is in no way the controlling factor as far as the shortage is concerned.

I think we are talking here about a shortage of petroleum and petroleum products. To me one factor is the shift away from coal: the increased cost of mining, the transportation problems, the environmental problems, and the stack gas problems that tended to shift central power stations to oil or gas. Power stations were shifting back and forth between oil and gas, and then they were frightened away from gas because of the potential shortage so they went to oil. This is one factor that contributed to the drainage of oil reserves.

Other usage factors include the increased number of cars purchased, the lower gas mileage on recent model automobiles, and then, of course, getting out of phase somewhat with the seasons. Last year we didn't have enough heating oil for the winter, so we had to run like crazy to catch up with heating oil, which helped cause a gasoline shortage for this summer.

I think it is important to remember that when we talk about the energy crisis, we are generally talking about finished petroleum products, and that we are talking about a 3 to 5 percent gasoline shortage last June compared to a predicted shortage of, say, 5 to 7 percent this winter. This situation is going to get worse, but I think it is relevant that the difference between supply and demand is still pretty narrow.

CONGRESSMAN UDALL: We live in a complicated society, and I want to underline what Congressman McCormack said. Americans like to point to the obvious villain, and there is always somebody—the dirty oil company or the environmentalists, or this politician, or that so and so who did such and such. The fact is that the environmentalists didn't cause the oil shortage and did not stop atomic power from coming on the line.

Russell Train, chairman of the President's Council on Environmental Quality, recently pointed out that fifteen years ago the average car got twenty-one to twenty-two miles to the gallon but now it gets only nine or ten. Well, maybe we have lost two of those ten or eleven miles per gallon because of the new clean air quality requirements. But the rest of the eight or nine miles per gallon are due to Americans' insisting on bigger cars with more options, like air conditioning and all of the rest.

Environmentalists have been pretty tough on new atomic plants, but you can't blame them for massive power shortages or for the fact that 50 percent of this country's electric power is not being produced by atomic plants.

MR. SPAHR: Mr. Udall, I think I mostly agree with you, but when we say we can't blame the environmentalists, I don't think we have to go further and say we can't blame environmental considerations for some of the things that have happened to us. This year, for example, we are selling a fair amount of distillate fuel to power companies that were using coal a year ago. This has added to the load that we are carrying, compounding the problem caused by the automobile, and it is a result of environmental considerations. We could cite other examples. So environmental considerations have had an impact on the problem.

CONGRESSMAN UDALL: Why, sure, I concede that, and I don't want to monopolize the time, but let me make a related point.

There is a feeling around the country today that this great environmental movement of the late 1960s and early 1970s was some sort of a fraud, that it wasn't really real. Well the fact is that the dangers—the things that led us into this environmental movement—are very real and very genuine. Our rivers are polluted. The air in New York will kill you. We are dumping waste in the oceans and the balance of life is endangered. We can't just give up on this environmental fight.

The aim has got to be abundant energy, but also preservation of the environment. If we suddenly decide that

8

the environmental movement was a fraud, we will be in real serious trouble.

SENATOR HANSEN: If I could add just a word here. I think that it ought to be pointed out that there is plenty of fault to go around. Basically we are all environmentalists. There isn't anyone who doesn't subscribe to the idea that the environmentalists' goals are laudable and worthwhile. But let's not bend over backward too far and refuse to assign a fair share of blame where it ought really to go. By that I mean that there really isn't any basic shortage of all forms of energy in this country. We have coal running out of our ears. Up in my state of Wyoming there is enough coal to last this nation for 300 years.

DR. McCRACKEN: I believe that is low sulfur coal.

SENATOR HANSEN: Low sulfur coal—and for reasons that we all know and appreciate, we have decided that it would be desirable to burn cleaner fuel. And the Federal Power Commission has contributed to the overall problem. It decided that it was in the public interest to regulate the price of natural gas, so it put an artificially low lid on the price of natural gas. Those persons who would like to blame the coal industry today for failing to rise to the occasion forget one basic fact: this is a nation that still embraces the capitalistic system, and I thank God that it does, because I think that really is the epitome of freedom. People don't necessarily decide that they are going to do what may serve the public interest unless it happens also to serve their personal interest. I think that is a great system, and I don't depreciate it one bit. The fact is, we have made it not very profitable for the energy companies to develop these other forms of energy.

As Congressman Udall and Congressman McCormack both have observed, the push was for certain kinds of energy. About 78 percent of all of our energy comes from petroleum, and now we are in a tight spot. Considerations of national security make us question if we ought further to pursue a policy that envisages greater dependency upon foreign sources of supply.

Our friends in Israel are well aware that the Arabs are flexing their muscles with the intent of persuading America that it had better listen to the interests of the Arab countries.

So, in a word, I would like to say that environmental considerations have played a part. They have tended to discourage the Department of Interior from leasing blocks of the Outer Continental Shelf where the prospects for oil are very good, and from leasing Western lands as well. The department said a year or two ago, "We want an environmental impact statement before you can drill a single well." As you know, Mr. Spahr, that was countermanded later.

These are some of the reasons that we are in our present fix. I think that we are getting out of it, but it is going to take a while. We won't do it quickly, and we are going to find that we can't waste energy as we have in the past.

DR. McCRACKEN: I guess the problem has been to try to achieve the proper balance in our concerns about energy, about environment, and about other matters.

Mo, I like your comment about pointing the finger. We do have a little tendency to be much more interested in the question, who is the villain, than in the question, what is the problem; and we do need to remind ourselves that it is the second question that we need to keep in focus here.

CONGRESSMAN McCORMACK: May I comment on that? So far as the environment or environmental protection is concerned, we should be asking ourselves: what are the trade-offs between the energy we need and the environment that we want to protect and do our existing laws serve that purpose?

I cite as a simple example the standards that have been set for sulfur dioxide emission—standards which simply preclude the burning of most of the coal in this country by central power stations. I am sure it was assumed that when those standards were set new technology would make it possible to successfully scrub the gas and take out the sulfur dioxide. Unfortunately the assumption was made without any technological assessment of the question. Today, because there is no scrubber technology in place or even in sight to do the job, we are not burning sulfur coal in the central power stations. The question is whether those standards are proper, or whether we should be talking about an emission level that we can reach with the scrubber technology we have (which would be ten times better than the level possible without any scrubber technology at all).

SENATOR HANSEN: If I could add just a word there, I think that the question is not whether we have some good scrubbers in existence today, but whether we want them to be perfect. If we insist that they must be perfect, that they take every speck of the sulfur dioxide out, then obviously we have got a lot of work left to do.

Instead of talking only about trade-offs we should also consider turn-offs. If we want perfection today we are going to have a lot of turn-offs, a lot of lights going out. And so I would hope that we can take a reasonable position.

CONGRESSMAN UDALL: Let me plug in on that. I hope Americans don't make one mistake that I see coming. For decades now we have worshipped technology. There has always been the hope that you can get a quick technological fix, like the addict, that technology is going to save you. But I am not sure in this case that technology is going to come up with all of the answers very quickly or very soon.

Let me underline something that was said here a moment ago. We ought to be talking right now—this fall, this winter—about reducing energy demand. We talk about supplies, whereas we ought to be talking about energy conservation. You can look all over this land tonight and see neon signs—in Las Vegas, Times Square, all around. We are going to have to put first things first—which is what this priorities argument is about. Do we want to have neon signs or do we want to have enough energy so that farmers can get their crops in next spring? Do we want to use the oil and the energy to take a 5,000-mile vacation this summer with the children, or do we want to use the oil to get people to work and keep some factories operating?

We ought to have in place, but we don't, an energy conservation program with priorities so that, when we get in trouble—and it is right ahead—we know what things are really important and what things are secondary.

MR. SPAHR: I agree, conservation is necessary—and immediately. Just as urgent is the need to do some of the things required to ensure that we'll be able to recover our position for the future. Both the short term and the long term are, to me, very serious problems. No matter how well we do between now and next March—and I think we should do the best we can—we are still going to be in at least intermittent trouble.

DR. McCRACKEN: What do you mean by that? Looking down the road, say, this winter and perhaps next summer, what is the ordinary family facing?

MR. SPAHR: Well, the ordinary family in one part of the country faces a problem entirely different from that family in another part of the country. The family in New England may very well be concerned about whether or not its fuel supplier will be able to keep the oil tank full, or perhaps more realistically keep it from going empty. The family down in Texas won't worry about heating oil, and I don't think it will have to be concerned too much about gas this year either.

So, different problems will exist in different places. Frankly, I am sure that all the oil companies are going to do their level best to make heating oil for homes their top priority. Of equal priority, I think, is fuel for farmers.

Crops are pretty important. If the farmer doesn't have fuel for his tractor at the right time, we are not going to get the product we need.

CONGRESSMAN UDALL: That leads me to two related points that I hope we won't lose sight of in these difficult months and years ahead.

The first one is that we are one country. It would really be outrageous if I were to sit in Arizona or Cliff Hansen in Wyoming, fat and happy with plenty of heating oil and gas, while some businessman has to close down in New England, or some hospital goes cold in Massachusetts. We are one country, and we have got to have a system that is fair to everybody; if we are going to have shortages, we ought to share them.

Second, we may get the idea in these difficult years ahead that the United States can be an island, that we can barter with the Mideast countries for our oil supplies and start bidding against Japan and our friends in the Common Market. All the industrialized nations are in this thing together, and we will make a very great mistake if we try to go it alone and outbid the democracies, our friends in Europe and in Japan, in an ever-tightening oil situation.

MR. SPAHR: To continue, if I might for just a moment, Mr. Udall. I really believe that the homeowner in general is going to get through the winter all right, because of what I have said about priorities. I am more concerned about interruptions of industrial activity due to shortages of gas and distillate fuel. This means that, here and there, people will be out of work and production will be interrupted—which isn't going to help our steady progress toward a higher gross national product.

That is what I think we will experience to some degree this winter, and to a greater degree next winter, because refining capacity is limited and cannot be increased by next year. We will not begin to get increased production out of the commitments that were made this spring until the latter part of 1975 or early 1976.

DR. McCRACKEN: So the cutting edge, as it were, of this problem will be in the industrial area, because factories, especially if we have particularly cold weather, will be unable to get gas and heating oil and that sort of thing?

MR. SPAHR: That is right.

SENATOR HANSEN: Well, without looking too far down the road, Dr. McCracken, I think it is important to think about some of the things we could be doing now to make the picture brighter. I'm referring to freeing the oil from stripper wells. For those who may not know, a stripper well is a well that produces ten or fewer barrels of oil per day.

There are about 350,000 stripper wells in this country, and they account for about an eighth of our total oil reserves in this country. If we have around 40 billion barrels of known oil reserves, and if we could increase the squeezing of those sands so as to get 1 percent more from the stripper wells, it would be like finding a 400-million-barrel oil field. Now that is pretty important. And unless we change what we are doing now, we are going to lose the production from these stripper wells. Any time that an oil man doesn't get a few more dollars from his stripper well than he is spending to retrieve the oil, he will plug it, and when a well is plugged it is lost for all time.

So, what we ought to do immediately is to exempt the oil production from stripper wells from all price controls in order to make certain that we get all of that oil. We can talk about the inflationary impact of removing these controls and all of that sort of thing, but the high prices in America today for oil and gasoline products are a reflection not of costs in this country, but rather of higher prices abroad.

In America domestic crude oil sells for less than imported crude oil. For that reason, one of the steps we could take now that would make a difference would be to let price be a factor in rationing demand. Higher prices would bring about the sort of economizing that Mo Udall speaks about. I agree with him 100 percent: we have

14

been a profligate nation in the use of energy.

We can't afford to waste energy any longer. I can think of no better way to stop that waste than to make energy more costly. And if we do that, we will also give incentive to the industry to take steps that can turn this picture around so that we won't have to anticipate shortages from now until the crack of doom.

DR. McCRACKEN: Yes, particularly since the margin here, apparently, is not great. I think, Mr. McCormack, you said that 5 percent or something like that—

CONGRESSMAN McCORMACK: That is right.

DR. McCRACKEN:—is a reasonable estimate of what the gap is now. Do you have any comments on this?

CONGRESSMAN McCORMACK: Only the reminder that the 5 to 7 percent that we are short this year will be substantially bigger next year—

DR. McCRACKEN: I agree.

CONGRESSMAN McCORMACK:—and that some of the solutions we are talking about apply only to this year. We talk about conserving energy. We can introduce all sorts of conservation programs and they will help us this year, but we can't continue to run solely on conservation programs. I agree we should institute conservation programs. But if we cut our consumption by 50 percent and do nothing about supply, eight years from now we will be right back where we are today, just because of normal growth.

We must conserve in every way we can. We have to remember, however, that there will be no new refineries on the line for about three more years. Demand, barring whatever conservation programs we can institute, will continue to grow, but supply will grow only to the extent we can increase production from existing facilities. So, the situation is going to get substantially worse. And one of the real perils we face is having two consecutive seasons of extreme weather, a very cold winter followed by a very hot summer, or vice versa.

CONGRESSMAN UDALL: I wanted to respond to what Mr. Spahr said a moment ago. I grew up in an era in which part of the great American myth was that everything is going to get bigger and better. If you have two cars, your children are going to have four cars. If you have a four-room house, they are going to have an eight-room house. Everything is going to get bigger and better. Mr. Spahr spoke of an ever-increasing gross national product. I think the day is going to come, and this may sound un-American, when we are going to have to ask ourselves whether we can really afford, in terms of energy, the belief that everything should get bigger and better all of the time.

Japan—and that is a pretty nice country—uses one-fifth of the electricity per capita that we use. Britain and France consume only about half the energy we do, and I don't think we are twice as happy as they are. It may be a tough adjustment, but I think we are going to have to ask ourselves whether we can go on and on.

The fact is that we will never again produce as much natural gas annually as we have produced in the past years. That is inevitable, whatever we do about pricing—and I am for some changes in pricing.

Our domestic oil production has also probably peaked regardless of all the drilling offshore and all the measures we want to take. We have to face the fact that either we have to find long-range new sources of non-polluting energy, or we are going to have to adjust the great American attitude that somehow we are entitled to double our energy production in the next ten or fifteen years. We can't increase production to this extent. Those days are gone. I regret it, but this is a hard fact that the American people haven't faced yet.

DR. McCRACKEN: Looking beyond next summer at the actions which can be taken to bring supply and demand

16

into some kind of balance, and recognizing, of course, that we want to work on the demand side by emphasizing conservation, what can be done about augmenting our supplies? What do you see there? One of the obvious questions that always comes up when we take a little longer range view is what can we expect from nuclear power? Obviously, reactors can't be built right away. But will nuclear power start to be a significant factor, say, by the end of the decade?

CONGRESSMAN McCORMACK: Yes, nuclear power represents about 1 percent of our total energy today. It will rise to 5 or 10 percent of the total by the end of the decade and to 35 to 40 percent by the end of the century. We may even find ourselves—because of new techniques in the nuclear business that we aren't even aware of yet— depending more on nuclear-produced electricity in the mid-1980s than we presently anticipate.

But it is slow. There is an eight to ten year lead time between the commitment to build a nuclear reactor and the time you get it on the line, and we are hoping to cut that down to five or seven years. I think we have about thirty reactors on the line now. We will have about 200 on line by 1983. But that is still fairly slow and even then, nuclear energy will constitute only 10 to 15 percent of our total energy.

DR. McCRACKEN: Are we going to run into a fuel problem there?

CONGRESSMAN McCORMACK: Not really.

DR. McCRACKEN: We seem to run into fuel problems everywhere else.

CONGRESSMAN McCORMACK: As a matter of fact, I think this is one area where the situation may turn out to be brighter than we now anticipate. There has been a great deal of fear about inadequate uranium supplies, but this fear is based on $8 a pound uranium. When we consider the whole cost of building a nuclear reactor, fuel is such an insignificant portion of the entire cost that we could have $25 or $40 uranium and it would make little difference. Besides, we are going to have the breeder program coming in and it will help us with our fuels. So, we will have adequate fuel. This is not really a problem.

DR. McCRACKEN: What is the breeder program? Could you describe that for a layman in a sentence?

CONGRESSMAN McCORMACK: Okay. The breeder reactor is simply a new type of nuclear reactor which produces more fuel than it consumes. Today, the nuclear reactors that burn uranium-235 produce plutonium, but they don't produce as many plutonium atoms as they consume of uranium atoms. For every ten uranium-235 atoms they consume, they produce six to eight plutonium atoms.

With the breeder reactor, we push that ratio over the top and make more plutonium than we consume of uranium, and thereby we are making more fuel than we consume. We will have our first demonstration breeder plant operative in the early 1980s and there will be other breeders: a liquid-metal fast breeder and a gas-cooled breeder will be operating on a demonstration basis in the 1980s. By the time they are really needed to make an impact on fuel supply—the late 1980s or early 1990s— they will be operating and providing the fuel we need.

DR. McCRACKEN: It sounds like perpetual motion.

CONGRESSMAN McCORMACK: Well, it isn't really, because you are consuming uranium-238 instead of uranium-235.

CONGRESSMAN UDALL: Well, I don't want to be the pessimist on the program. But let's assume that you get all the atomic energy you want and that Congressman Mc-Cormack's programs are successful, as we all hope they are, and that we can quadruple the amount of energy we are consuming in this country. There will still be a number of problems. What are we going to do with the atomic waste? What will the impact of the greater consumption be on the environment? Where are we going to dump the hot water that is produced as a result of all these atomic plants? What about the possible change of climate in our cities when all of this electricity and energy is being produced and consumed?

Americans have always been behind the times in terms of assessing the impact of technology. We look down the road and solve one problem, but sometimes find we have created two or three more. I don't want to overemphasize it, but I still think Americans have got to find the time to back off and say: We are not going to double our energy consumption every ten or twelve years. Enough is enough.

CONGRESSMAN McCORMACK: Mo, I agree with you. I don't believe we are going to go on doubling our consumption every eight, ten or even every twelve years. I think we are going to level off. This is inevitable. And I think we will do it partially by design. For one thing, we will run out of oil and natural gas and will have to replace those fuels with liquefied and gasified coal—which is a clumsy solution, I think, to the problem. The days of unlimited gasoline use for individual automobiles is already on the decline. I expect that we will be looking back on this time and finding that we lived at the peak of the golden era of consuming fossil fuels in the 1950s and the 1960s and that we are now on the other side of the slope. I expect that we will undertake programs to reduce that consumption—one of which will be to reduce the heat dumping in cities. I even think we are going to deliberately set out to redeploy our population.

CONGRESSMAN UDALL: The small towns are coming back.

CONGRESSMAN McCORMACK: The small town is coming back, and the small model village—where you may not need or even be allowed to operate a private automobile—is also coming back. There are all kinds of schemes today for forbidding the use of private cars in central cities. This concept of devising environments in which the private automobile is unnecessary or forbidden is going to grow. One very logical solution is the redeployment of population under the land-use plan that you and Senator Jackson are working on—so that we will have new model villages and won't have to use automobiles.

But let me speak to the other half of your comment just for a second—the storage and management of radioactive waste and the release of excess heat into the atmosphere. We are obviously not going to dump excess heat into our waters. There are very few bodies of water in this country that could handle the heat from a thousand-megawatt power plant. The Columbia River could and some of the Great Lakes could under limited circumstances. The rest of the waste will go into the ocean or into the air. From a meteorological standpoint this is no problem at all, except in our large metropolitan areas such as the Los Angeles Basin or the Boston-Washington strip. We have testimony stating that, at the present growth rate in energy consumption, the temperature in the Los Angeles Basin will jump fifteen degrees Fahrenheit by 1990, just from the increased consumption of energy, and the temperature in the Baltimore-Washington strip, six degrees Fahrenheit.

These are the things, along with the reduction in the availability of gas, that are going to drive us into model villages and away from the use of the car.

DR. McCRACKEN: What that would mean, then, is that we would have to have more power for air conditioners, and therefore we would need more power.

CONGRESSMAN McCORMACK: No, I don't think so. I think there is a lot to be done in terms of how to use energy in buildings and how to design buildings. That is a whole additional area.

Let me respond, if I may, to one of the questions raised earlier. Mo has brought up a point we both recognize as a bugaboo among some members of the public, that is, the storage of radioactive waste. Radioactive waste storage is a simple matter of good management and good engineering. The United States has an excellent record in this area; we know what we are doing, so this presents, as far as I am concerned, no real problem.

There is no question about the fact that we have to store radioactive waste. For the first time in history, we have the problem of providing honest-to-God perpetual care, perpetual storage—because we are going to have to take care of those wastes forever. But this can be done in a very small amount of space. We could do it on the surface—in concrete buildings, if nothing else—and there are a lot of concrete buildings standing around today that will handle radioactive waste. The process buildings of the Manhattan project and subsequent AEC projects will handle all of the waste we produce for the next 100 years.

MR. SPAHR: I want to agree with Mr. Udall that we cannot go on increasing our consumption of energy at the same annual rate that we have been experiencing for the last several years. If we do this—and most estimates still put the normal projected rate of increase at 4 percent per year—we will double our present consumption in sixteen years. As all these gentlemen know, that is an impossible thing to do.

I believe the now rather famous National Petroleum Council study that was submitted to the secretary of the interior several years ago indicated that in fifteen years from the date of that study our imbalance of payments arising from oil importations alone would be about $30 billion a year. That projection was based upon the experience with oil prices up to that time. The increase in prices has been so great in the last six months—I am referring to what the Middle Eastern nations have done —that the $30 billion gap will come a lot sooner on a lot less oil than was imagined.

So, the pressure on us to find ways to conserve energy is much higher than I think any of us on this platform imagined a year ago.

21

So the challenge is clear. We have to do the obvious thing now—conserve—while working on the more difficult long-range projects for the time that Mr. McCormack is concerned with.

We emphasize the things that are bad in this country all too much while overlooking the things that are done pretty well. I know this is a controversial statement, but I will submit it anyway. I think that we are far enough advanced technologically, notwithstanding all the bad experiences that can be cited on undersea work, to support a decision to go ahead with offshore exploration and development in the areas of the East and the West coast where geological surveys suggest great amounts of oil exist. I think oil can be discovered and developed in these areas with a rather minimum amount of trouble.

We ought to develop those areas, we ought to get the oil out of Alaska, we ought to use coal where coal can be used without real danger to the population and, at the same time, we ought to continue the research Mr. McCormack is concerned with so that we will have different sources of energy ready ten years from now.

In my view, if we emphasize the positive in this country and take a few minimal risks we can solve a good part of our problem.

SENATOR HANSEN: If I could just add a little postscript to that. I couldn't agree more. I think that we need to keep in mind that the real energy crunch in this country will come within the next ten years, and it will come while we are still largely dependent upon oil and petroleum products. We can get on with the technology of coal gasification, coal liquefication, oil shale development, uranium or breeder reactors, solar power, all of these things—but these developments won't be on line making a significant contribution short of ten to fifteen years from now.

So the real focus must be on the next ten years. First of all, let's eliminate every unnecessary use of energy. I don't mean by that to give some energy czar in this town the authority, arbitrarily, to say—Mo, you are using too much energy in Arizona, and Cliff, you are using too much in Wyoming; rather, we would let prices do that—let price be the controlling factor.

Second, if we can explore those untapped resources now—remembering that the undiscovered oil and gas lying on the outer continental shelf and within this country probably amounts to 100 times the amount of oil and gas used in 1971—we can ease this crunch. This would give us the lead time necessary to get on with coal gasification, shale oil development, and the other opportunities available to us to shore up our resources.

But I think we must realize that a lot of steps will have to be taken in the next ten years and certainly among those is the encouragement that most other nations recognize the oil industry needs if it is to discover and produce oil from these untapped resources that we do not now have at our disposal.

CONGRESSMAN UDALL: I don't want to disagree with Senator Hansen's general thesis but, you know, we had better face the fact that sooner or later mankind will be out of oil and gas.

Putting myself out in the future, as if I were a grandfather holding my grandson on my knee, I can imagine the boy saying—"Tell me, granddad, about those crazy old days when electrical utility companies used to advertise to get people to buy more appliances and burn more electric heat, and when gas stations would give away glasses and stamps to get you to use more gas. Tell me about all those strange things that used to happen."

If you take a pessimistic view, mankind will be out of oil and gas in twenty years. If you take an optimistic view and accept Senator Hansen's theory, there is a whole pool of oil and gas offshore and we can drill for it and we won't run out for maybe fifty years. In either case, we had better be looking down the road and making plans for the long haul.

We must also be deciding what amount of environmental damage we are willing to accept. My subcommittee in the Congress is now considering strip mining legislation. How much of Montana and Wyoming and Appalachia are we ready to tear up to maintain our present level of consumption? How many oil spills are we ready to risk? How many beaches will be spoiled, how much marine life wrecked, if we calculate wrongly in all of this offshore drilling? I am sure some of it can be done safely. But we are going to be balancing a lot of factors as we go down the road.

SENATOR HANSEN: I think the Coast Guard has made some important observations. Exactly one-and-a-half percent of all the oil pollution that occurred in the oceans of the world in 1971 came from offshore drilling. Far more of it came from cleaning out tankers and spills and that sort of thing—by bringing it over from the Middle East. More than 30 percent comes from automobile crankcase drainings running into sewers and ultimately into the ocean.

If I may make one other observation, I don't mean to say, Mo, that the world's petroleum resources are inexhaustible. I do say, though, that it is important that we take advantage of what resources we have to help get us through this crunch until we're successful with coal gasification, until we get oil shale, until we get houses properly insulated, until we get uranium and perhaps solar energy working for us. Let's take advantage of what we now have, because our first priority should be to make certain that we keep all of our options open as a nation.

I don't want somebody else telling us what our foreign policy is going to be or how many people are going to go to work in this country. I want people working, I want jobs, and I think that we can do lots of things. We shouldn't act as if everybody has it made, as if it's unimportant whether the neighbor down the street goes to work tomorrow. It is important, and I want to see to it that we have enough energy to get us over this tough spot.

DR. McCRACKEN: It seems to me the emphasis that has come out of this discussion is summed up in this word "balance." We aren't looking at a future where there will

24

be no gasoline, no electricity, no energy. Rather we are probably moving into an era where energy supplies will no longer be so abundant and so cheap that we really don't have to think about them—which has been the situation for the last two or three decades. All of us are going to be hearing a lot more about new ways of generating energy—about fast breeder reactors, the liquefaction of coal, and the like—and about the importance of conserving energy, because energy will be in more limited supply relative to demand. That means, and we should face up to it, that energy is probably going to be far more costly in the period ahead than it has been during much of our lifetime.

CONGRESSMAN UDALL: It should be.

DR. McCRACKEN: And of course, Cliff, you emphasized that higher prices for energy are one of the most effective ways of encouraging people to economize, and I suppose to stimulate supply, too.

CONGRESSMAN UDALL: I come from the Southwest where the sun shines all the time, and I am very interested in solar energy. When I ask why it is that we can't go ahead with solar energy, I'm told that it is off in the next century somewhere, that technology has to be developed. But there are something like two million solar water heaters in operation in Japan today. One can go into a Japanese hardware store and buy a little gadget to put on the roof that will heat water. In some places in America, in some months, half of all the energy use of the household is due to heating water for the dishwasher, the washing machine, the shower and all the rest. Yet, with gas and oil so "cheap"—and I put cheap in quotes—we haven't really used solar energy because it was more "expensive"—and I put expensive in quotes. We are now going to have to look at some of these alternative sources and face the fact that energy isn't running out of our ears any more. From here on out we have to be careful of the way we use it.

MR. SPAHR: Dr. McCracken, I do think we have to think of challenging concepts. To be less than serious for a moment, I wonder how much power we could save in this country if we were to revert to radio communication and turn off these high temperature lights that we use on TV appearances. Do you think that would be an acceptable idea to all these gentlemen on this panel?

CONGRESSMAN UDALL: Never to congressmen. It would be a body blow to the whole political establishment. [Laughter.]

D R. McCRACKEN: The time has come now for questions and comments from our panel of distinguished guests. Does someone have a question?

GORDON SUMNER, Office of International Security Affairs, Department of Defense: I would like to address my questions to Congressman Udall and to Senator Hansen. They have both alluded to the international political implications of the energy crisis, so I would like to ask them to focus a little more on the global, strategic implications.

The best information we have suggests that Japan is importing about 85 percent of its oil from the Middle East. Italy imports about the same share, 85 percent, the United Kingdom about 84 percent, and West Germany 75 percent. Given that these economies are so intertwined, what will be the strategic implications for the United States as we are confronted with a worldwide energy shortage, with maybe fifteen or twenty years of vulnerability and stress? Will we be judged able to meet the situation with balance and to see our way through the very difficult period ahead?

CONGRESSMAN UDALL: Well, I would start off with this: There is an old church hymn, "Count Your Blessings, Name Them One By One." Of all the major industrial nations, there are only two today that are self-sufficient in energy. One is the Soviet Union and the other is Canada.

We in the United States are concerned because we have to import all this oil. Well, the fact is, we are much better off than the British or the French or the Germans or the Italians or any other country—particularly the Japanese. So we should consider ourselves fortunate. We have a little more margin than some and a lot more than others.

But, to reiterate a point I made earlier, it would be a grave mistake for us to go it alone and to say: we come first; the Germans, Japanese, British, are our friends, but we are going to take care of ourselves first. That would only bid up the oil prices in the international market and open possibilities for those who do not wish us well. I hope we will have a free-world global strategy on energy, that we will put our heads together with Germany, Japan, and the Common Market countries in order to look ahead and plan, at the same time that we are trying to make ourselves self-sufficient just as soon as we can.

To the extent that we depend on foreign sources, particularly unstable foreign sources, we could be vulnerable some time down the road to international blackmail and all kinds of foreign policy pressures from other countries. So we had better get on with both of these things at the same time.

SENATOR HANSEN: I would say that first of all, it is important to understand who is bidding up the prices of whose oil. The United States hasn't been bidding up foreign oil. I don't mean to imply that we aren't interested in the price of oil or that we don't need a lot of oil, because we are now importing a little over a third of all the oil that we are using. At the present time we are using about 18 million barrels a day, about a third of which comes from Canada, Venezuela, or the Middle East, and to a lesser degree, from other areas.

Now, Japan has been actively bidding for oil. It has been using the dollars we have been paying for Japanese goods and services and has been actively engaging on an individual basis—not in collaboration with the other countries, with the OECD nations, which includes Japan, but pretty much on a unilateral basis—in bidding up prices for oil. It has also been coming into the United States and buying coal property. And I don't think the Arabs are reflecting a situation of competition between oil consuming and oil importing nations, so much as they are taking advantage of a situation with which they are intimately familiar—that is, that a lot of nations have to import their energy requirements. The Arabs have been expropriating properties, and insisting that the price of crude oil rise. They have been out in the forefront.

I think that what Mo suggests makes good sense. The oil consuming countries should get together, and I hope that we can. But Japan might be reluctant because it knows perfectly well that, having to import 98 percent of its energy, it has got to get oil wherever it can.

We certainly can become very susceptible to blackmail, to attempts by oil exporting countries to change our foreign policy. This would not only hurt us, but it would also be a disservice to freedom-loving people the world over. The best thing that America can do, it seems to me, is to take all possible steps to lessen our dependency upon foreign supplies.

Just two years ago, the Canadian premier, Mr. Trudeau, said, "Don't ever be concerned about the dependability of supplies from Canada." But lately he has been saying that we are not going to get much more oil. Canada was selling more than half of the oil it produced a couple of years ago and buying more than half of what it consumed. Now it is talking about a trans-Canada pipeline and the need to keep oil at home.

ARNOLD PACKER, Committee for Economic Development: Congressman Udall, I noticed that there is considerable sympathy with your suggestion for more conservation. Was there sympathy for conservation when Congress considered the highway bill last year? Also, if that bill were brought up this year, would Congress be so willing to spend that money for highways?

CONGRESSMAN UDALL: No, it wouldn't, because Americans react to crises. We're not always ready to take the painful and unpopular steps until the crunch and the squeeze comes on. I was one of those who helped lead the fight to bust the Highway Trust Fund. We were clobbered. But still we won a modest sort of victory: We were told that three years down the road we would get a little piece of the highway fund to use on mass transit. But we are going to have to move very quickly in diverting funds into the mass transit area.

I personally believe—in line with some of the suggestions that have been made here tonight—that American fuel prices have been unrealistically low. The British and the Europeans have had 80- and 90-cent gasoline for a long, long time; we have had 30-cent gasoline. I would favor putting on a 10- or 12-cent gasoline tax per gallon right now, in order to raise funds for mass transit to subsidize subways and bus services. Because we are going to have to get people out of their private automobiles in the next decade and into transit modes that use a fifth or a tenth or a fifteenth of the amount of energy that my little car takes.

NED BEACH, Senate Republican Policy Committee: I am addressing my question to Mr. Spahr and Congressman McCormack. I am wondering whether we could handle the problem of conservation in a different way—namely, reduce the expenditure of energy by causing the users to absorb more of the energy requirements than they have in the past. For example, we heat our houses too hot in the winter and we cool them too much in the summer; men wear heavy clothes in the office and the ladies are cold, so then they heat the place up. In other words, the wrong thing is controlling what we do.

Has any thought been given to ways in which we can put the responsibility on the consumer to provide more of the energy that he is himself requiring? For example, we could ride bicycles, set thermostats lower, or change men's fashion to shirts without ties in the summer. That would save a lot of money and a lot of energy.

DR. McCRACKEN: And maybe red flannels in the winter time.

CONGRESSMAN UDALL: That would put a lot of people in the clothing business out of work—which is part of the trade-off.

MR. SPAHR: Well, Mr. Beach, some thought has been given to the kind of problem you have described. I know of a number of oil companies that are well along in programs to demonstrate conservation. They are finding ways to reduce power consumption in their own refineries beyond what they used before, and they are discovering that they can, in some instances, achieve a power reduction of as much as 7 or 8 percent.

It is through savings on the part of industry that we will be able to achieve the greatest part of the conservation we are looking for. I don't mean to say that it isn't important for each of us to get a certain personal sense of self-discipline as people. Lowering thermostats, that will help. But the greatest possibilities are going to come through conservation in industrial use. Once we are in a position to demonstrate what can be done with facts and figures, we should have a rather powerful argument for other industrialists to do the same.

However, I don't think the user in every instance, particularly an industrial user, is in the driver's seat. My own company is a user as well as a producer of energy because we make, among other things, a nitrogen chemical for fertilizer. We have been severely limited in our purchase of natural gas, which, as you know, is a major raw material for that kind of product. The only way we can make it up is by using certain refinery streams, which are also in short supply. If we do that, then we have to cut back on our production of distillate. So, we are finding that we have to do less than we could if we could run at full capacity, and others are going to find the same kind of problem confronting them.

CONGRESSMAN McCORMACK: I agree that we have developed a society and a social ethic that calls for the exorbitant consumption of energy.

You know, out in the future this world is going to look back on this century as one grand orgy of burning fossil fuels. It started at the beginning of the twentieth century, and it will finish at about the end, when we run out with a great big "bleep." We should be aware of the fact that we are less than thirty years away from the probable exhaustion, for all practical purposes, of massive quantities of fossil fuel. I do think that conservation programs are a very substantial part of the answer, particularly in the short run.

Let me point out that the Northwest—Washington and Oregon and Idaho, the Bonneville Power System—is facing the most severe shortage of electricity in its history. Rainfall is the lowest it's been in ninety-five years, and 90 percent of the area's energy is hydroelectric, the rest being nuclear and coal. So you can see what kind of a situation we are in out there.

The Washington legislature has just given the governor virtually dictatorial powers to enforce conservation measures, and the governor of Oregon has, I believe, assumed similar authority on his own. We will have a good case to study as far as conservation is concerned, because the Northwest has to cut its consumption of electricity by at least 10 percent before the end of 1973. And this is going to be an across-the-board cut, including commercial, industrial and domestic consumption. I think there will be a big market for 25 and 40 watt light bulbs in the Northwest in the next couple of months.

Now, I also agree with you about our dress. We do overheat our houses in the winter time and overcool them in the summer time. When Lou Roddis, the president of Consolidated Edison, came to testify before my subcommittee, he said: "People want to know what we can do now to save energy. I tell them that I have instructed all the male animals—and I am paraphrasing him—in our organization to wear sport shirts without neckties to work, and no coats." He was sitting there testifying before a congressional committee without a necktie on. When I congratulated him for that, he replied, "Well, it is pretty lonesome doing this all by myself." So I took off my necktie as well.

We can use less electricity in heating and cooling our houses.

MR. BEACH: Thinking of such ideas as that, if all the telephones in the country were made with a very large dialing system that was hooked up to the power plants so that when you dialed the number you also put some electricity in the system, it would force the user to supply some of the power that he uses. [Laughter.]

DR. McCRACKEN: Could we have another question?

ARTHUR I. MENDOLIA, Assistant Secretary of Defense (I&L): I would like to address my question to Congressman Udall.

The discussion here has concentrated on short-term answers to our problems. But even if this country were to be extraordinarily successful with conservation, it would only postpone for perhaps a decade the day when we would have to rely on new sources of energy. I think Dr. McCracken would agree that in general the consumption of energy is directly related to standard of living. The United States, with 6 percent of the world's population, uses a third of the world's energy. Shouldn't we adopt public policies that accept that as a fact and direct our attention and our resources to generating new sources of energy so that the world may benefit?

CONGRESSMAN UDALL: Well, I am all for new, clean, non-polluting, permanent renewable sources of energy, but I am not sure we are going to be all that happy if we quadruple our energy sources. If in the year 2000 homes are four times as big as they are now and the average American is consuming four times as much energy, I don't think we would be that much happier or that much better off.

And we would raise a lot of very serious questions about what we're doing to the environment if we take that route.

Have I answered your question?

MR. MENDOLIA: No, because I believe you are taking the American view. My question is—whose standard of living are we going to freeze? I pointed out that Americans use a third of the world's energy and yet we are only 6 percent of the population. What about the rest of the world? Are we going to ask the rest of the world to freeze its standard of living for the next several hundred years at a level equal to that of the lower third in the United States?

CONGRESSMAN UDALL: No, no, I wouldn't buy that at all. There are people on the bottom of the economic scale in this country that deserve a bigger share of our energy consumption. And there are billions of people in this world who have a right to get above the level of subsistence living. We ought to set a standard of restraint, thrift, and conservation for ourselves, without saying to the less fortunate people of the world that they have to stop where they are, too.

Wealth is not very well distributed in this country or in the world. I don't want to sound like a socialist, for I believe in the enterprise system, but there has to be a better allocation of energy consumption in the world than there is today.

CONGRESSMAN McCORMACK: May I comment? I think we must place a heavy emphasis on research and demonstration programs in order to get new sources of energy on the line as rapidly as possible. It should be a primary element in our national energy policy to have solar and geothermal energy demonstration units functioning as quickly as feasible—during this decade if possible and certainly within ten years. As far as fusion is concerned, we should have demonstration fusion reactors on the line by the mid-1990s.

We should be able to export this technology, not only for the production of energy or the conversion of energy

33

to useful forms, but also for the inexpensive transmission of energy and for maximum efficiency in consumption. I'm referring to fuel cells, heat pumps, batteries, this sort of thing. Research and development programs in these various fields should be totalling at least a couple of billion dollars a year. And we should be exporting the results to all the world. Energy technology is one of the prime exports we can produce, and it has the advantage of reducing or mitigating the dependence of one nation upon another as far as energy is concerned. For example, when Japan has a sound nuclear energy base, it is not going to be dependent upon the Middle East. And the same thing applies to us and other nations as well.

If we can look forward to a twenty-first century with fusion energy available to all nations in the world, then we will have taken a major step away from the problems that have caused wars in the past. I think this should be a major element of our energy policy.

MR. SPAHR: Let me add something to that question also. For the next ten years I don't think the choice is going to be ours. It will be the producing countries that will determine whether there is going to be an increase in power utilization in the undeveloped countries vis-à-vis the United States. We are not going to have the privilege of making that decision. And I also think it is almost beyond question that we are going to have to face up to the problem of temporarily reducing our own standard of living to some degree.

DR. McCRACKEN: Or at least limit the rise.

MR. SPAHR: Or at least limit the rise, but I really think it is going to amount to more than that, Paul.

SENATOR HANSEN: If I could just make one observation, it would be to say that that was a great speech you made, Mac, and I appreciate it. But I don't entirely agree with you. Without throwing cold water on your ideas about how to bring world peace, let me say that I don't see any correlation between a rise in the standard of living on the one hand and peaceful pursuits and goals on the other. If you are saying that we can bring about peace by making energy abundant the world over, I would be skeptical—although always hopeful.

It occurs to me that all too often Americans embark upon worthy goals, and certainly I endorse the twin goals of abundant energy and world peace. But maybe we would make a pretty good contribution if we were to demonstrate anew to the world the importance of everybody working, of taking responsibility, of trying to mind his own business, and of not telling everybody else around the world how to live, how much energy to use, and so on. I'm reminded of some of the counterproductive efforts we have made. We have gone abroad with lifesaving techniques, with penicillin and surgical techniques—and what has happened? We have saved people from dying from disease and so they now starve because there isn't enough wheat to go around.

And I don't have any answers. Your idea, I think, is one that most Christians would like to embrace. They would like to share with the world. But I don't necessarily think, being very cruel about it, brutally frank, that it always works that way.

CONGRESSMAN McCORMACK: May I just say that I don't agree. I hold my position that to the degree any individual nation is freed from its dependence upon some other nation for energy, to that degree some of the tensions and some of the pressures that may lead to war have been relieved. This is what I am getting at. And if we develop this technology, it is just a matter of making it available to those countries that want to pick it up. Certainly, then, I think we have provided a major service, in our own self-interest as well as in theirs.

LEONARD LANE, Friends of the Earth: I would like to speak to the point made by both Congressman Udall and Senator Hansen that the price mechanism must be used as the primary device, or as a primary device, in achieving better energy conservation. It would seem to me that if that is the case, doesn't it follow that we should question some of the tax subsidies and direct subsidies we have been granting the energy industry? That is to say, if we are in fact in danger of running short of fossil fuels by the end of this century, does it make sense to continue to allow percentage depletion allowance right off of intangible drilling costs, the foreign tax credit and so forth, all of which in fact reduce the market price of energy, and therefore encourage its excessive consumption?

SENATOR HANSEN: Let me take a crack at the question first and then give you a chance to rebut—which you all do very well. I would have to say that we did that several years ago when we passed the 1969 tax reform law. We reduced the depletion allowance from 27.5 percent to 22 percent. This saved the government—or nicked the oil companies, however you want to regard it—about $600 to $700 million a year.

Just within the last month, Chase Manhattan Bank revised upward its estimate of the balance-of-payments deficit we will be facing by 1976. Earlier this year it had estimated that in 1976 we would be paying out $17.5 billion for energy bought abroad from foreign countries. A few weeks ago it increased that estimate to $19.5 billion— a reflection of what the Arabs have done in changing the price of crude oil.

Now, that is the economics of it. If you want to visit more of that kind of saving on American taxpayers, you are welcome. For my part, I don't think we need that kind of help. I would rather not have to put out $19.5 billion for $6 and $7 a barrel foreign crude when, with a little more incentive to the domestic industry, we could do a far better job here.

Let me make one more point, and that is that people seem to think of the oil industry as a big, fat-cat unit. They don't differentiate between the majors, those companies with operations in the Middle East or Indonesia and elsewhere, and the independent driller. It is the independent in this country who has discovered about 75 to 80 percent of all our onshore discoveries. He is the guy who drills forty dry holes to find the one producer, the guy who drills about seventy holes before he gets one that is *really* an economic producer. This has been the track record on wildcatting in the past. When we reduced the depletion allowance, it was the independents who had second thoughts about staying in the oil business. Many of them went into cattle feeding and into any other number of things that looked at the time like better ways to make money than to drill holes in the ground.

We need to give this segment of the industry greater encouragement. And obviously if we are concerned about inflation, it would seem to me that the best insurance we can buy against higher prices is to encourage domestic industry to find oil and gas by drilling deeper, more expensive wells, and to let foreign suppliers know that they do not have as much of a monopolistic hold on the market as they would like to have.

CONGRESSMAN UDALL: One of the country's great economists, our Paul McCracken here, will be comforted to know that I really believe in the market system, in market incentives and the value of pricing mechanism. And one of the great energy executives in our country, Mr. Spahr, will be pained to know that I think the oil companies have committed many sins and that not all of their past policies and practices have been in the interest of this country.

I happen to think these depletion allowances, and a whole bundle of other tax benefits we have given oil companies over the years, have been an outrage. Yes, we have to give incentives for more oil production, gas production, drilling, and this sort of thing, but we don't have to make hundreds of millionaires in Dallas and Houston in order to encourage greater oil and gas production.

I do think that we are going to have to review our energy policies, with an eye to keeping those that encourage thrift and greater production and discarding those that don't. We must look at every aspect of this thing. What a crazy system we have had all these years—a kind of "drain America first" approach! The oil companies fought for years to keep out foreign imports so that we could burn up U.S. oil first; and now we turn completely around.

Getting back to the market mechanism, one thing we can do is make it more expensive to be profligate in the use of energy. Today, if you and I both have houses, and if I run my air conditioners full tilt all summer and overheat the house in the winter and have every appliance known to man, and if you are very thrifty, I am rewarded and you are not. If I use twice as much energy as you do, I am rewarded by a lower rate on that second half. The situation ought to be reversed. We should have one energy rate that covers the homeowner's basic allocation, enough fuel to heat and cool the house to a reasonable degree, and a second, *much* higher rate that applies to everything above the basic allocation. And we ought to do the same with industry.

In this way we could use the market mechanism to encourage thrift.

SENATOR HANSEN: Dr. McCracken, like my wife, I want to have the last word. When the Six-Day War broke out, this country had the productive capability to satisfy its own needs and those of all its allies throughout the world. It doesn't have that capability now. In working out new energy policies, we shouldn't forget for one minute the importance of being as nearly self-sufficient as we can.

DR. McCRACKEN: Charley, do you want to make a comment on that?

MR. SPAHR: Thank you very much, Dr. McCracken. First, a note of levity. I wonder if Mr. Udall has in mind that if his thesis is carried out, all the big homes of the future would be built in the Southwest instead of in New England. It just could work out that way.

CONGRESSMAN UDALL: No, no. No, no.

MR. SPAHR: Second, I believe I can speak to the problem with a bit of objectivity. My company is a semi-integrated oil company that never has been able to produce enough oil from its own wells to feed its refineries. At the

38

current time, we produce 13 percent of the oil we run and buy the rest. Three years ago we were able to buy almost all of what we needed from domestic sources. We now buy 40 percent of it offshore—and next year, 1974, the share will probably be greater. So, when I talk about an oil company that is a producer, I think that you might agree that I can look at it objectively, not with bias because of a desire to protect myself.

Oil companies in general do not have a better return on total capital employed than all manufacturers. In fact, for the last several years, I think their return has run a fraction of a percent less than the average. So, if their profits, including windfalls and tax savings and all that, are compared with the profits of other manufacturers, I wonder what the argument would be? They are not making an unconscionable return, in my view, particularly when it is realized that they plow their capital back in the ground to find more oil and other energy sources.

In addition, the oil companies are going to have an awfully difficult time in the next decade raising money to do the total job that has to be done, because their cash generation will simply not be sufficient. They are going to have to borrow more and more money and, to do that, will have to demonstrate an adequate rate of return.

So much for my opinion on the adequacy of the tax allowances for the oil industry.

DR. McCRACKEN: We have explored the tax question and other aspects of the energy problem very thoroughly here tonight, but now we must draw this AEI Round Table to a close.

The thrust of the evening's discussion has been that the energy problem must be approached from both ends. We must think in terms of new supplies and new sources of energy. We also must recognize that the old bargain basement days when we could buy unlimited quantities of energy at very low prices are over. Americans are going to have to become more conscious of the need to be prudent in their use of electricity, gasoline, fuel oil, or whatever it may be.

Let me thank our panel members—Senator Clifford Hansen, Congressman Mike McCormack, Congressman Morris Udall, and Mr. Charles Spahr of the Standard Oil Company of Ohio—for a lively discussion of an urgent issue. [Applause.]

PART II

FUTURE OPTIONS FOR ENERGY POLICIES

Paul W. McCracken, *Moderator*

⟡

Jennings Randolph
Mark O. Hatfield
Dixy Lee Ray
Philip H. Trezise

PAUL W. McCRACKEN, Edmund Ezra Day university professor of business administration, University of Michigan, and Round Table moderator: Through most of our lives we Americans have been used to driving into the filling station and saying "fill-her-up," or turning on the air conditioner when we have been hot and raising the thermostat when we have been cold. The idea that there might not be anything at the other end to respond to our command really never occurred to us. Then, all at once, we've been having brownouts in the summer and we find ourselves threatened with cold houses and gasoline shortages this winter. To the ordinary person, this problem seems to have hit and to have become urgent very suddenly.

The focus for this evening's discussion is future options for energy policies. But, first, a little background. Senator Randolph, you have been following energy policy for a long period of time. How did we get to where we are today?

JENNINGS RANDOLPH, United States Senate (Democrat, West Virginia): Well, Paul, we had an energy shortage back in the 1940s. I had the responsibility at that time of cosponsoring with Senator Joseph O'Mahoney of Wyoming the Synthetic Liquid Fuels Act, and I remember so well the opposition we had. We wanted to process coal into aviation fuel (as the Germans had already done successfully). We wanted to extract oil from oil shale, and we wanted coal gasification and other possibilities.

We started some good programs and pilot projects

were under way. Then the submarine menace subsided, the crisis was over and the American people said, "Let well enough alone." So the funding stopped, as you and others know. It is almost tragic that it did because we were ready to move forward, as we should have then and in the 1950s, in doing this job.

I remember offering resolutions in the Senate in 1959 to establish a joint committee on energy and in 1970 to establish a national fuels and energy policy commission. No one was really interested in the ideas and so they died. But we have kept at it and at it. And now, for the last two-and-a-half years the Senate's national fuels and energy policy has been operating under a Senate resolution that Senator Jackson and I and over fifty others have fostered. The Interior Committee, the Commerce Committee, the Public Works Committee, and the Joint Committee on Atomic Energy, under the study, have been really digging into this problem. Now because we are faced with growth of population, with our increasing reliance upon petroleum from countries that may not be as friendly tomorrow as they are today, the problem has become more serious. This time if the flow of oil is stopped, it won't be because of a submarine menace. It will be because of political decisions by the oil-exporting countries.

The Senate's National Fuels and Energy Policy Study has already been presenting some substance in the way of legislation, and we have other measures that we think will be very helpful. I will not go into them now. Perhaps we will talk about them a little later.

I think it is important for us to realize that so often in this country we act after the fact rather than before it. But what's frustrating in this case is that we had the facts long ago which pointed the way to what should have been done. Some thirty years ago I flew in a plane fueled with gasoline processed from coal—West Virginia coal processed into aviation fuel.

DR. McCRACKEN: In other words, you flew in a coal-burning plane?

44

SENATOR RANDOLPH: Yes, from Morgantown, West Virginia, to here—to National Airport. Senator O'Mahoney and I were attempting to dramatize the need for the legislation we were supporting. One doesn't feel too confident riding over the Allegheny Mountains for the first time, you know, with that kind of fuel. We made it, but we were a little squeamish, I must say. [Laughter.]

Of course, Dr. Ray wouldn't have been worried, because she is so courageous. She comes before the National Fuels and Energy Policy Study and speaks with such fluency and expertise.

DIXY LEE RAY, chairman, Atomic Energy Commission: I was just going to ask, was it the plane you were worried about or the fuel in the plane?

SENATOR RANDOLPH: The fuel in the plane, I must admit, Dr. Ray.

DR. RAY: I think that we ought to recognize here that your foresight and that of other members of the Congress, is well known. And there have been people with foresight in various agencies too. The Geological Survey has done studies to show the nature and the extent of this country's renewable fuel resources. More recently, the Office of Science and Technology undertook a very important study on energy problems. And a number of universities—Massachusetts Institute of Technology and Columbia University among them—have had important energy studies.

So, various public and private institutions, as well as the fuel companies themselves, have been warning us that a shortage was coming about. Thoughtful people have known for quite a long time that we were headed this way, and there was going to come a time when energy supply would be limited.

But as you say, we really have to be right at the edge of danger or crisis before the threat gets into the consciousness of the average person.

SENATOR RANDOLPH: Yes, Dr. Ray, I think that is true. If a person can't go to work, he will realize there is a problem, won't he? If his house is cold or the hospital is cold, he will realize it. Such things make an impact upon people.

DR. McCRACKEN: Well, there is no way we can turn history back and do the things that it would have been desirable to have done earlier. What should we do now, Mark? What do you see down the way now?

MARK O. HATFIELD, United States Senate (Republican, Oregon): Paul, I think we ought to think in terms of a short term and a long term.

First, we have the traditional energy sources upon which we can base new technology for more efficient use of what we are actually producing now. Let me give you an illustration. The great grid system of the Columbia and Snake Rivers produces approximately 10 million kilowatts of hydro energy. With the projects that are authorized and those that are under way, we can double the output of existing facilities without building new dams. In other words, we can put a third generator at Grand Coulee, a second power house at Bonneville, and so forth, so that we can make the existing systems more efficient.

Secondly, there is intriguing potential in what might be called the exotic future—in nuclear fusion, geothermal energy, solar energy, and so on. Here in Oregon we are working on the possibility, through hydrogenation, of translating wood waste into low-sulfur oil. There are about 55 million tons of wood waste accumulating every year in this country. If we recovered just 10 percent of it, we could produce 7 million barrels of oil.

Let me add one further thing that I think shows the interrelatedness of the various aspects of the energy problem. We cannot just address ourselves to existing or future sources of energy without biting the bullet on some very difficult decisions. Since World War II, we have not only added 60 million people to our population, but we have trebled our consumption of energy. This problem of huge demand is not due to electric toothbrushes alone, and it won't be solved by the simplistic approach of just turning off lights in the rooms we are not using. That is good energy conservation, but it won't solve the total problem.

We have made technological choices that have been counterproductive—counterecological and counter the requirements of energy policy. For example, we have moved from natural fibers to synthetic fibers and from forest products to aluminum, light metals, plastics, and glass—all of which demand more energy. I am not anti-light metals, but let me remind you that the production of aluminum, cement and chemicals consumes about 30 percent of the industrial use of electrical power in the United States.

We have a new law in Oregon prohibiting the use of nonreturnable cans and bottles in the soft drink and beer industries. It has been computed that if the nation-at-large had this kind of law, we would be able to light the cities of Washington, Boston, San Francisco and Philadelphia for a six-month period just with the energy we would save in that one area of technological production.

These are some of the related problems that we are going to have to address. The decisions won't be easy. But I don't think we can go on our merry way, expecting to find new exotic sources or improve existing sources while refusing to address ourselves to these technological production questions. The choice between the forest product that grows by free solar energy and the one pound of aluminum that's produced with 30,000 BTUs is but one example of these technological choices I think we have to make.

SENATOR RANDOLPH: Paul, I don't want to interrupt, because perhaps Phil Trezise wants to speak, but I want to direct a thought to all of you—

DR. McCRACKEN: Go right ahead.

SENATOR RANDOLPH: —and especially to you, Mark, because you talked along this line.

Personally I want the Alaskan pipeline very much. I want it to be constructed and used. But people are thinking of it as a final answer, a panacea. Yet even with all the petroleum that would come out of the Alaskan venture, as highly successful as we would want it to be, we would only be providing about 10 percent of the needed petroleum supply of the lower forty-eight American states. In other words, as has been said in other ways here, there is not just one answer to our energy problem.

SENATOR HATFIELD: Yes, certainly.

PHILIP H. TREZISE, Brookings Institution: Paul, I think Senator Hatfield made a useful distinction which I would like to follow up, and that is the distinction between the short term and the long term—and incidentally it applies to the pipeline, I might say. In the short term, it seems to me, there isn't a great deal that can be done, because the policy options are very limited. Consumption trends are fixed, and they will not change quickly. Supply adjustments will take a long time in most cases, even in the best of circumstances.

So, we have to face a couple of years—between now, say, and 1976—in which I think we are going to have higher prices for energy, and that is that.

SENATOR RANDOLPH: The consumer is going to pay.

DR. TREZISE: Yes, the consumer is going to pay.

Now, I think there are two useful things we could do in the short run. One would be to decontrol the price of gasoline. In my view, it is highly inadvisable to have a freeze on gasoline prices if we are indeed short of gasoline. Another would be to deregulate natural gas. To me it is scandalous that the country should be paying discount prices for a premium fuel in a time of fuel crisis.

These two steps could help a good deal in the short run, in terms of bringing about needed adjustments in consumption and in the allocation of resources. But basically the energy problem is a short-term and a medium-term problem, whereas most of the solutions, I am afraid, have a long-term aspect to them.

48

SENATOR HATFIELD: May I make another suggestion on that point?

DR. McCRACKEN: Yes.

SENATOR HATFIELD: We have a rate structure today that rewards consumption—the more fuel or electricity you consume, the lower your rates. Reversing this is another immediate short-term project we could undertake. But it's a tough one because people are used to the current rate structure. I think we would find that some very interesting conservation practices would occur voluntarily and very quickly if we eliminated this incentive to consume.

DR. McCRACKEN: If I may, as an economist, step into the discussion here—I am supposed to be the moderator—

SENATOR HATFIELD: You have the right to an opinion, Paul.

DR. McCRACKEN: Well, I have a good many of them. [Laughter.]

SENATOR RANDOLPH: An expert has been characterized as a person who is seldom right, but never in doubt. [Laughter.]

DR. McCRACKEN: You know, no better way has been found for encouraging people to economize on something that we need to be more prudent with than to have them recognize it is going to cost more. That is the point Phil was making here. We do, in a sense, have the paradox of the federal government encouraging the use of gasoline by a bargain basement price at a time when we're running short.

Well, what about it? We will want to spend a good deal of time here discussing the fundamental, long-term issues, but what kinds of things can the ordinary person expect for the next twelve months or so? What is a reasonable expectation?

SENATOR RANDOLPH: I can't tell you, of course, except for fragments of the situation. I know that the Senate has passed a measure providing for mandatory fuel allocations, which I believe in—and I hope the House will act on this bill very, very promptly.

Second, we also passed a "sense-of-the-Senate" resolution, supporting a reduction of speed limits on interstate highways by ten miles an hour. Now, that might not seem very important, but it would amount to significant savings in gasoline. I addressed a letter supporting this to the governors of all the states and I am delighted to say, as I did in the Senate today, that more than twenty-eight have replied affirmatively that they are moving to do just this.

This is just one little step that we can take that would be helpful. We certainly are not going to have fewer cars operating, despite what people say. Highways will be built, cars will operate, trucks will move, and buses will be on the road. This is a fact of life.

DR. McCRACKEN: I am sure you are right.

SENATOR HATFIELD: I think in the next year we will really not save so much as we will become more conscious of what our limitations are, and perhaps learn conservation. As the senator points out, a lower speed limit may save only a little, but if you combine that with the possibility, say, of addressing ourselves to home insulation, the savings begin to mount. Studies have shown that about $250 in insulation added to the 1,500 square foot home of average construction would cut the fuel bill in half. Up to now we have given very little attention to the requirements in our building codes that address themselves to this kind of conservation.

I think these are examples of what we are going to learn. We will have to learn to do without, we will have to learn inconveniences, and we will have to become more aware of limitations and perhaps more willing, then, to accept some of these higher costs and to support the idea of additional tax monies for energy research.

And now the Senate Interior Committee would designate your agency, Dr. Ray, as a lead agency in a new commitment of R & D. Hopefully, the House will go along with it, we will fund it, and OMB [Office of Management and Budget] will not impound the monies after we've appropriated them—so that we can get on with some of these projects, both those with short-term, and with long-term impact.

SENATOR RANDOLPH: I hesitate to speak again because Phil, I know, is over here ready to comment, but—

SENATOR HATFIELD: We have no time limitations. The other two are just going to have to interrupt. [Laughter.]

DR. McCRACKEN: Go ahead, senator.

SENATOR RANDOLPH: As examples of what we can do, several scheduled airlines in the United States have already reduced the speeds of their aircraft, with a considerable saving of fuel. As you know, Mark, in the last few days we have given attention to this in the Senate— asking the Air Force to see if it can't draw back and conserve fuel by similar methods.

DR. McCRACKEN: Phil, why not solve the problem by just importing more oil and gas?

DR. TREZISE: Well, at the moment, importing more isn't a highly promising solution, because tankers are hard to come by and also, indeed, one of the big suppliers is apparently cutting back on output.

DR. McCRACKEN: Which supplier do you mean there?

DR. TREZISE: Libya.
So, I don't think in the short run we can expect much of an increase in supplies. In any event, we don't have the refineries to process a great deal more crude oil, even if we could get it.
Really the situation is pretty much fixed for the short run. All the things we are talking about are great ideas—

but for some future period. I don't see much for the next couple of years but higher prices and a certain amount of rationing through the price mechanism. If I were advising the Senate, I would suggest a small hike in the excise tax on gasoline. That would be about as salutary a thing as could be done.

DR. McCRACKEN: At this point in our discussion, let's look down the way a bit. Apparently in the short run we are going to have to make do with heavy emphasis on conservation, slower speeds, better insulation of homes and the like. But looking further into the future, how about solving the problem with nuclear energy?

DR. RAY: I think there are so many fine possibilities when we get beyond the short term that the real problem is to select the kind of thing we ought to concentrate on, and then to marshal the resources in trained manpower, scientific talent, and financial support and to fix the responsibility for getting the job done.

I have to agree with all of you as to the immediate future. Really, it is set. We have to make do with what we have—with the supplies that we know are available—and there will probably be shortages. There will have to be mandatory allotments, and hopefully they will be set as fairly as possible.

But looking further ahead, we don't really have either an energy or a fuel problem at all—if we want to use the resources that we do have. And your state of West Virginia is a good example, Senator Randolph.

The United States has lots of coal. Most of the coal of the North American continent is located within our borders. If we wanted to concentrate on coal alone, we could derive a sufficient amount of energy from our coal resources to last us at least for several hundred years, given today's rates of utilization and a reasonable increase.

But there are problems, many problems, and we recognize them very easily.

First, how do we get the coal out of the ground without running head-on into the environmental problem of

52

the scarring of the land and the difficult problems of worker safety? Some say we can use the coal without bringing it out of the ground, which means *in situ* gasification, or retorting it under the ground. But that has its own kinds of problems.

Second, in the transportation and combustion of coal there are problems of the ash, of clinkers, of waste products, and of the effluents. This comes as a surprise to a lot of people, but the fact is that, without treatment, more radioactivity is released into the air from the burning of coal than from a nuclear power plant.

DR. McCRACKEN: How is that again?

DR. RAY: Yes, sir. There is natural radioactivity in most rock and stone, and in the earth itself. When you burn coal, the amount of radioactivity from the natural radioactivity of the coal gets into the smoke and, unless you have a special device to filter it out, there is a greater release of radioactivity in the effluent of burning coal than is permitted in a nuclear power plant—because the latter has special devices to filter it out.

SENATOR RANDOLPH: But coal can be made into clean synthetic fuels.

DR. RAY: Yes, it can be made clean, exactly.

SENATOR RANDOLPH: And that is important.

DR. RAY: Right, I agree with you completely, sir.

SENATOR RANDOLPH: Spoken like an Appalachian. [Laughter.]

DR. RAY: Thank you. There is coal in my part of the country, too. There's coal in the West, in the Southwest, in the Northeast, and here in the Central Atlantic area, and so on.

We have the special problems of low BTU coal, of low-sulfur and high-sulfur coal. All of these varieties require special processing, either to make the coal clean before it is burned, or to clean it up in the process. Also, there is the option of conversion, or hydrogenation: by adding hydrogen, coal can be liquefied. So, a whole spectrum of liquid fuels, including high-test aviation gasoline, can be made from coal; or by going further, coal can be turned into a product which is essentially natural gas.

What all of these things require primarily is engineering development, not so much basic research as engineering development, to work out the details of the process applicable to each kind of coal. The process that might work with high-sulfur Appalachian coal would not work so well with the low-sulfur Western coal, and so on.

We need, I think—I am using just one example here —a tremendous extension of the program already begun in the Office of Coal Research, a reasonable sized pilot plant to test out engineering developments, to test out a variety of processes in order to learn what one is going to be the best.

This does take time—five to ten years. But the option is there to be used, and that is just one example.

DR. McCRACKEN: May I raise a question? There is, of course, the problem of air pollution—the sulfur and that sort of thing. But there is also the problem of strip mining—in other words, the problem of getting at the coal.

SENATOR RANDOLPH: You are correct, Paul. And surface mining today accounts for more than 50 percent of all U.S. coal supplies moving into use in our economy. So it is a real problem. Mark is very familiar with the Senate's efforts to bring a sensible bill on this to the floor. I hope we will bring out a well-reasoned bill, one that will stipulate high enough penalties so that no one will surface mine and then move on to another location without restoring the land. Small fines won't work; the fines must be very high and the penalties must be written into law.

We can rehabilitate the land as we mine it. This has

been proven over and over again. We must do now what we failed to do a long time ago. In the past, the land was desecrated, no doubt about it. But from here on out we now can do what is necessary in the way of information.

For many, many years surface mining will be a real source of the supply of coal in this country. So I hope, Mark, the right kind of bill will come out of the Congress, one which represents a balancing between energy and environmental policies.

SENATOR HATFIELD: What Jennings has said illustrates, I think, another overriding issue that politicians particularly have to face up to and assume responsibility for—that is, to reduce the polarization that exists today between economic need and environmental values. Too often we have seen politicians who have exploited this polarization for the emotional value of votes at the ballot box, rather than trying to reason, and trying to bring the various interest groups together.

I am fairly persuaded, from my eight years as chief executive of my state at a time when our major thrust was economic diversification and development, that political leaders can be the catalysts for cooperation. Paralleling our efforts to enact more laws governing air and water pollution, to build more parks, and to express more environmental concerns in law, we showed that there was not a mutual antagonism—that, with political leadership, we could serve economic needs without violating environmental values.

But I think this goes beyond the engineers and the researchers. It requires a political leadership that is responsible in trying to be a catalyst for agreement rather than an exploiter of the emotional issues and divisiveness we now have in this country.

SENATOR RANDOLPH: Mark, I like you. You know, you vote in the morning a certain way, and in the afternoon when the issue is before the Senate, you vote just as you did in the morning. I like that, indeed I do. [Laughter.]

SENATOR HATFIELD: I don't know if there is any virtue in consistency.

DR. TREZISE: These are all questions of cost in the end, aren't they? I saw some figures this morning on the costs of reclaiming Western coal land. The numbers run from, say, 2 cents a ton to 16 cents a ton—according to the findings I saw, and apparently they are substantiated by a certain amount of background material.

Now, I suppose in West Virginia the reclamation of strip mining land may be more expensive—

SENATOR RANDOLPH: Yes, it is, because of the terrain.

DR. TREZISE: —but in the end, if we want coal as a substitute for oil, we are going to have to pay for it.

DR. McCRACKEN: It seems to me you are saying something very important. What concerns a great many people is that the coal may be there, but they don't want to ruin the landscape by surface or strip mining. What seems to be being said here is that this kind of mining can now take place on an economic basis with a restoration of the terrain.

Is that a fair statement?

SENATOR RANDOLPH: Yes, I believe that can be done —and I am an environmentalist and I have worked for all the laws that we have for air and water pollution control, solid waste disposal, and the whole scope of environmental policies.

But in a time like this we have to be realists, and not so much advocates for environmentalism or industry. We have to be realists—and we can be. There is no need to have polarization, as you mentioned, Mark. There can be a partnership and common goals. This is important. Scare stories won't do anyone any good.

I would like to talk just a moment about the coal situation. I gathered certain figures this afternoon that I think, Paul, might be helpful to your viewers and audience. For fiscal year 1974 the Nixon administration requested a total of $778 million for energy research programs. Of that amount, $115.9 million was for coal, but $556 million was for nuclear fission and fusion. Now just think of that! What happened?

DR. RAY: We have to be more persuasive.

SENATOR RANDOLPH: Just look at that situation. The request for nuclear research was four times greater than that for coal-related research. Admittedly the President recommended in June an increase of about $100 million for energy research. Some $50 million of this was already appropriated by the Congress for coal research. Nevertheless, there is a great imbalance here, isn't there?

DR. RAY: Yes, there has been, and I think that it has come about because apparently, until recently, we haven't felt the necessity for development research in these other fields.

SENATOR RANDOLPH: That is right.

DR. RAY: I think we are going to see, and I certainly hope so, a much better redressing of the imbalance.
But there is another reason for the large figure for nuclear research. At the basic scientific research stage, when you are studying whether a process is going to work or not, when you are involved in the laboratory and in certain field experiments, the costs are not exorbitant. But once the basic research is done, when you start to build a demonstration plant or when you are developing the engineering to prove out a process on a large scale, that is when the big costs come. It happens that in the nuclear field right now, we are at that stage of a big demonstration project. That is where the majority of the funds actually are.

DR. McCRACKEN: You are talking about the new breeder, the fast-breeder reactor?

DR. RAY: That is right. And I think and hope we are going to see a similar kind of increase in funding very soon, not only in the coal field, but also for the other energy options that you referred to a few moments ago.

SENATOR RANDOLPH: Such as geothermal?

DR. RAY: Yes, geothermal. There is really a piddling amount of money going into research on geothermal energy right now. But just as soon as we want to start pilot projects, as we are hoping to do soon, to develop the hot-rock techniques that have been pretty well tested in the laboratory, to develop the hot brine which is available in the Southwest, and to develop more of the dry-seam areas, then we will be talking about enormous sums of money—because we will have to go out with drilling rigs and build a big installation to bring up that steam or that heat and turn it into electricity.

SENATOR RANDOLPH: Dr. Ray, what about the gas-cooled reactor? I noticed it is funded at $7.9 million. What do you see here?

DR. RAY: The gas-cooled reactor is an alternative to the light-water-cooled reactor. It has been in development for some time. The Atomic Energy Commission has done quite a bit of basic research in this field.

But the main development in commercial-size plants has been made by the General Atomic Company. The first of these plants, the one at Fort St. Vrain, Colorado, is going to come on line very shortly. It is just now being tested and will come up to power very shortly. In contrast, the Peach Bottom (Pennsylvania) plant, which I think you all know about, is a research operation designed for further work on the fuel elements themselves and on the fuel processing and so on.

The advantage of the gas-cooled reactor is the possibility of its use in places where water for cooling purposes is in short supply—that is, in arid regions or in places where pressures on the water supply may be so great one does not want to make use of it.

DR. McCRACKEN: It seems to me that the problem that really bothers the ordinary person about nuclear energy is the very simple question—is it safe? We still hear a great deal of discussion about this and a fair amount of skepticism. What do you say to that?

DR. RAY: If these plants were not safe, the Atomic Energy Commission would not license them, would not permit them to be built, and certainly would not permit them to operate.

There are two things I would like to emphasize here. First of all, there is no easier way to scare anybody than to say "radioactivity." And I think that we in the field have to admit we have not done a good job of giving the public a basic understanding of what radioactivity involves.

SENATOR RANDOLPH: That is the point. You've been talking to yourselves.

DR. RAY: We have been talking to ourselves, you are quite right.

SENATOR HATFIELD: May I bring out a useful point of information by asking you a leading question?

DR. RAY: Yes.

SENATOR HATFIELD: Tell us about the Hanford experience that so recently raised this question in the minds of many people, not only in the Northwest, but also in the entire country. You have an excellent response and I would like to have you use this opportunity to express it.

DR. RAY: As I think everybody knows, we recently had a leak from the waste storage tank at Hanford, Washington. The leak involved 115,000 gallons—which sounds like a big number and it is. We are distressed by it, because nobody likes to have anything leak out, particularly a material that has a lot of radioactivity in it.

59

But I think we have to put the event into perspective. We should recognize, first of all, that the waste in storage at Hanford is waste that has accumulated over a period of thirty years, almost exclusively from the weapons program. In the process of producing material for the weapons program—which our country for thirty years has deemed to be necessary in our national security program—65 million gallons of waste material have been accumulated there.

The decision to bury these wastes in storage tanks in the ground was made nearly thirty years ago. I don't think that we today are in any position to point the finger of blame at the people who, in the light of the technology then known and with the best available information and intent, made that decision. Hanford was selected as a location for the production of plutonium and for the waste storage because it was remote and because the characteristics of the soil there were suitable for a receptacle for the waste. In fact, clearly, it was never expected that those tanks would hold the waste forever. It was expected and understood that, in time, in ten to twenty years, they would corrode and the waste would leak out into the ground, and the earth itself would be the final receptacle for the waste.

We no longer consider that a good way to handle it. But that was what did happen, and we are now stuck with doing something with that decision. Because that decision was made, however, and because it was expected that the soil would be the receptacle, that soil has been studied for thirty years. In fact, we know more about the soil characteristics and the geology of Hanford than of any other place in this nation. Over the years there have been some leaks. We know exactly where the radioactivity is; we know exactly what the characteristics of the soil are with respect to absorbing and holding it; and we know where it will move over what periods of time, and so on. So I can say without question that there is no danger.

But the stuff is there, and beginning in the late 1950s, it was decided that it was no longer a good scheme just to leave it there in liquid and allow it eventually to leak out into the soil. That decision led to a program, supported and funded by the Congress, to gradually turn these materials into solids.

It takes time to build the equipment to pump out this highly radioactive material, to move it into an evaporator plant—all this under such conditions that none of it gets out in the environment—to evaporate it and then to return it as essentially a salt cake, a solid, to the holding tank, because as a solid it won't leak out. The evaporators were built, and the process of evaporation was begun in 1965. So far about one-third of the wastes have been converted to the solid state. We are now accelerating the process by adding a new evaporator which will come on line early next year. At the present rate, with this new evaporator, all of the liquid will be converted into the solid state by 1976.

So, we feel we are really on top of it.

DR. McCRACKEN: Phil, we don't want to leave out the international aspects of our subject. You indicated that tankers are short now and so, in the very near term, more oil from abroad doesn't constitute a solution to the energy problem. But all projections seem to suggest that we are going to rely substantially more on foreign sources of oil in the future.

This gets us into foreign policy. What kinds of questions and possibilities do we need to be thinking about?

DR. TREZISE: I think that in the medium term, by which I mean the latter part of this decade, we are going to be in a somewhat different situation than—

DR. McCRACKEN: By the way, I think it is worth emphasizing that by medium term you mean toward the end of this decade.

DR. TREZISE: Right.

DR. McCRACKEN: In other words, this is a long lead-time problem.

DR. TREZISE: Right. Most of the things we have been talking about tonight—nuclear power, geothermal, all of them—are for the 1980s. They really are not practical possibilities for the 1970s.

But, by the latter part of this decade, it is a fairly good prospect that the high prices that will have obtained for the next few years will have reduced consumption, the growth of consumption, and stimulated some new supplies, for example, oil from Alaska, the North Sea, and so on. Thus there will be an easy period, relatively, when the world price of oil will be softer and the pressure on supplies will not be so great. That may be the most difficult period for deciding on policies, because all the long-range projects we've talked about tonight are costly and require long lead-time investment, because we will have to make choices among them, and because the viability of one or another is by no means proven. All of this is going to come up, it seems to me, at the end of the decade when the crisis will seem a good deal further off.

If we want to become self-sufficient, we will have to make some very hard decisions at a hard time.

DR. McCRACKEN: Well, let me ask a very blunt question. What does this mean for us in terms of the Middle East? That is an uneasy part of the world and it is where the oil is. How do we feel our way through this problem?

DR. TREZISE: Well, it seems to me there are two parts to the problem. One is purely emotional—or political, if you will—and it is not predictable. If there is another war in the Middle East, I don't know what will happen. I don't know how anybody knows what the producing states will do. There are the various Arab states, there is Iran, and they have differing interests and differing capabilities. But this is essentially beyond prediction, one of the hazards we have to live with.

The other is the bargaining position of the oil-producing states, and that, as I say, may not be so strong in the latter part of the decade as it is today. The market uncertainties they will face will be considerable. So it will be possible, it seems to me, leaving aside the unpredictable political events, to strike arrangements with them which will be tolerable. Thus if we do want to go to self-sufficiency in the longer term, or to a greater degree of self-sufficiency than ordinary trends would bring, we are going to have to make, as I say, some excruciatingly difficult decisions about spending a lot of money at a time when it doesn't seem very necessary to spend a lot of money.

DR. RAY: Let me add something that seems relevant here.

Most people don't realize that a good many nations in Europe and particularly Japan are converting to nuclear energy at a much faster rate than is the United States. What are they doing? They are buying American-designed nuclear power plants, and the fuel for these plants, which is enriched uranium, is essentially an American monopoly. We have a native technology which is very highly desired abroad. Right now the AEC has firm contracts for the sale of $20 billion worth of enriched uranium to foreign countries. This is a contribution to the balance of payments that I think few people really recognize.

SENATOR RANDOLPH: I'm interested in the possibility of deepwater ports. Will they help?

DR. TREZISE: Well, that situation is still another constraint on our ability to—

SENATOR RANDOLPH: Transport effectively.

DR. TREZISE: Yes, in the short run. Until we have deepwater ports, we can't take the big tankers.

SENATOR HATFIELD: I want to go back a bit because I don't think we can just pass over the Middle East situation with general phraseology about Arabs and Israelis. This is a very explosive political issue. Right or wrong, we can't fight the Six-Day War over. My point is simply this: the Nixon administration and the government of the United States have lost credibility with the Arab states. Any way we look at it, if we are to have any kind of guarantee of future oil supply, we have to implement an even-handed, balanced foreign policy in the Middle East.

Now, I suppose it is not very political to say these things. But we have to bite the bullet. We must recognize that Israel cannot have an unlimited call upon our military supplies based upon its own determination of parity or security. And as to the matter of a secure Israeli border, that border has to be secure on both sides, not just on one.

All I am saying is that we shouldn't be pro-Israel—as I think our policy has been—that we have lost credibility and support amongst many of the traditionally pro-Western Arab states, and that I don't think we have to be anti-Israel any more than anti-Arab. We should be pro-American—looking at ourselves as a peacemaker and having our oil supplies guaranteed because the Arabs have confidence and trust in our foreign policy.

I have said it and I am not sorry.

DR. RAY: Good for you.

DR. McCRACKEN: By way of summarizing our foregoing discussion, if I had to identify just the two or three points that seem to me most important, one would be the news that technology is developing so that we will be able to make use of our coal resources, with prudent regard for the environment. Another is your point, Dr. Ray—and you are certainly right—that if there is any word in the English language that scares us, sort of paralyzes us, it is "radioactivity"—but apparently nuclear energy is safer than many of us think. Third, I am interested in the point that although we face a medium-term problem, there may be a change in the balance between the demand and supply for oil in the longer term.

May we now have questions from the floor? Yes?

FREDERICK WEINHOLD, Energy Policy Project: We have talked about options and about the tough choices that will have to be made. But who should make these choices, and how should they be made? Specifically, where do we place the superports? Do we go into states like Colorado, Wyoming and Montana with massive coal developments even though some people say we might not be able to restore the environment? When we put in coal plants or make major environmental changes on the beach or coastline, who should make the decision—the local people or the federal government? Finally, how do we see to it that we get energy while meeting environmental and local needs?

SENATOR RANDOLPH: Let me attempt an answer. I think the Congress has the major responsibility for such decisions. I believe, however, that there should be partnership between the executive and the legislative branches of government.

The Congress has attempted to create such a partnership, but the executive branch was not interested. That is exactly what happened. We tried to bring the executive and legislative together behind the formation of a national fuels and energy policy for the United States many, many years ago. We tried it again three-and-a-half years ago when I proposed establishment of a joint legislative-executive branch commission for this purpose.

All I can say today is that there must be an understanding between the executive and legislative branches. We can't accomplish this without a working partnership.

This business of accusing someone, of pointing the finger at some agency or committee or individual will not suffice for the task ahead. We also are going to have to have a tremendous energy R & D program in the United States, amounting to some $20 billion over a ten-year period and going in several directions. To do this, I think we are going to have to put aside narrow political considerations, conflicts between the administration and the

Congress, between Republicans and Democrats. We have to rise above that sort of thing. We can't afford to do otherwise.

I noted you talking about Hanford, Dr. Ray—when did we pass that legislation in the Congress?

DR. RAY: 1946.

SENATOR RANDOLPH: Yes, Hanford came into being a long time ago. But what happened in my state of West Virginia? Why, I was almost submerged because I voted for Hanford. A competitive situation existed at the time of coal versus nuclear power.

We can't afford that limited view any more. Time has run out on us. That is why I think that now is the time for a working partnership between Congress and the executive on the development of alternatives.

DR. McCRACKEN: Mark, would you like to comment on this?

SENATOR HATFIELD: I think it is a difficult question to respond to because it includes both specifics and very broad generalities. Let me illustrate what I mean.

Recently we have seen some evidence of "balkaniza-tion" in this country. I heard the other day that three state governors have indicated that facilities in their states are not going to produce oil for consumption in other parts of the country. Now, of course we can use political clout to impose upon the energy producing states the will of the non-energy producing states (or the lesser energy producing states) because the latter have more votes. But we can't afford to permit that kind of a situation to develop, and that is going to take leadership.

I think Americans are willing to contribute to a common need if they feel that everyone else is con-tributing too, or if they feel available resources are being distributed justly. That means we are going to have to do something like what was done up in my part of the coun-try. The Pacific Northwest produces a surplus of cheap hy-

66

droelectric power, and California is an area of need. Some time ago an intertie was suggested, but many Oregonians and Washingtonians said they were not going to let California rob them of their power. Well, when we looked into it, we learned it could be a two-way flow—from the Northwest to steam generators in California and then, in case of need, back up to the Northwest—and we began to develop a public understanding that we are one nation, and we are even one region.

So, again, it is going to take leadership at the level of the central government, but with the full input of state and local governments and of private industry and other private parties. It is easier to outline than it is to implement, but I think it is an example of what I mean by "leadership."

JOHN HIGGINS, *Business Week*: I would like to follow up on that. Congress has before it now a proposal for an energy research and development administration. The prospects for passage looked good in the spring, when Senator Randolph and Senator Jackson were talking about funding of $20 billion. But now the bill seems to be bogged down in various sorts of bureaucratic infighting over who is going to spend the $20 billion—the Interior Department or the Atomic Energy Commission. Even the proposal for an interim agency to kind of give birth to the program seems to have gone by the boards, and everyone is arguing about who is going to get the money, who is going to have the prestige.

How can the Congress resolve this problem?

SENATOR RANDOLPH: Well, John, this is a real problem. In the Senate we have jurisdictional conflicts among committees affecting many legislative proposals. But in our ad hoc Senate study we have brought four committees into the picture and they have worked well together. I don't believe that legislation is dead; it will come into being in a matter of weeks. That is my feeling. And I don't think Congress will recess in October as some people have said.

You are correct about the squabbling and the infighting. But somehow or other we have to follow through, because it is necessary that we do so.

DR. RAY: I would just like to recall that the administration sent the proposal up to the Hill back in July—I think. This is September. Now, one of the purposes of the Congress is to look closely at legislation, to debate it and try to examine all the possible outcomes and consequences. No major piece of legislation, particularly a reorganization bill, ever gets acted on in just a few weeks' time. I would like to suggest that, important as the issue is and even though we feel the great necessity for quick decision, this is always the time to be rather prudent and slow and to be sure that all of the consequences of such a reorganization bill are carefully examined.

Furthermore, it is easy enough to lay something out that looks awfully good on paper. Dr. Gorman Smith, who happens to be sitting down there at the table, tells a nice story that is applicable here. He points out that when you put a thin sheet of clear acetate over a contour map, then that map offers no impediment to a wax pencil as you mark out a pathway across it, but the poor fellow who has to climb across that terrain may find some great difficulties. So, you can draw up a reorganization plan but all the little boxes on the paper don't tell you much about the problems you will face when you really start to figure out who is going to do what.

We at the AEC are particularly sensitive to this because the reorganization bill would require a major change in the structure of the Atomic Energy Commission. For the last several weeks, the AEC staff and commissioners have been considering just how our agency would be torn asunder, what would go into the nuclear energy commission, as it is proposed, and what would go into the energy R & D agency. We are talking about people, about breaking up teams that have been working together, and separating budgets and programs and responsibilities. These things are not easily done overnight.

I think I would agree that the legislation is by no means dead but it *is* going to be debated in depth and there will be many more hearings. Then something will come out.

SENATOR HATFIELD: May I quickly distinguish between the reorganization bill providing for a Department of Energy and Natural Resources and the energy research bill calling for $2 billion a year for the next ten years. When we introduced the research bill, authored by Senator Jackson, the only signal we got from the administration was zero. But now we have a very strong signal of $1 billion a year. We are almost ready to mark up that bill in the Senate Interior Committee, having nearly worked out the management structure for this research program.

So the idea of setting up an energy research program and funding it is not being neglected. In fact, it is being expedited in a very significant way.

SENATOR RANDOLPH: Mark, may I add that over thirty senators are sponsoring that legislation, so there is an input of senatorial strength there beyond just the committee alone.

SENATOR HATFIELD: And the management structure can fit into any reorganization decision that we may make later.

SENATOR RANDOLPH: That is right.

DR. McCRACKEN: Mark, you said you were about ready to mark up the bill. What do you mean by that?

SENATOR HATFIELD: It is a colloquialism, I guess. It means that the committee is almost ready to make final decisions on a proposal we have heard a lot of testimony about.

DR. McCRACKEN: It's about ready to come forth.

SENATOR HATFIELD: Yes. The gestation is just about finished, and the midwife has been called. [Laughter.]

GEORGE LENCZOWSKI, University of California at Berkeley: Listening to the panel, I could sense an interest in, or emphasis on, self-sufficiency and economic productivity. Isn't this contradictory to present-day trends in the world, trends especially since the Second World War? And perhaps particularly in connection with our current interest in expanding Soviet-American trade in technology and other goods, should we not maximize our international commercial opportunities with regard to energy by following policies that would remove artificial obstacles to trade?

DR. McCRACKEN: Phil, you were assistant secretary of state for economic affairs. I think that question is right in your bailiwick.

DR. TREZISE: I had hoped I was careful not to commit myself on whether we should aim for self-sufficiency or not. In my view this is an open question that needs a good deal of debate. There obviously is some considerable hazard in being dependent on other countries for so vital, so critical, a commodity as energy raw materials. That is not really an arguable question. On the other hand, we may be able to get our energy supplies a great deal more cheaply if we import them, and this may be true for a long period of time—or forever, for all I know.

I am not sure what the answer is. I do know that self-sufficiency will very probably be a costly affair. If we are going to be self-sufficient, the American people had better be told that they are going to have to pay for it.

DR. McCRACKEN: By "pay for it" you mean higher prices?

DR. TREZISE: Yes, higher prices.

SENATOR RANDOLPH: Could I supplement what Phil has said in answer to your very probing question?

70

We see Japan and European countries faced with the necessity of buying their energy supplies from other countries, and we know they have no alternative. But we also know that Russia and the United States possess, within their borders, the ability to be energy self-sufficient. I think it is incumbent on our country to get that job done just as quickly as possible.

I feel very strongly about this. Where the security of the United States is involved, I don't think we can always rely on supplies from the Arab countries, or African countries, or what not. What do you think, Mark?

SENATOR HATFIELD: Could I put this into a totally different context, perhaps a more sensitive one. We represent 6 percent of the world's population, and last year we consumed 40 percent of the world's energy.

It seems to me we should look at this in a moral or ethical context. In my view, we in the United States have a responsibility not only to continue to upgrade our own standard of living—our way of life, comforts and whatever other terms you want to put it in—but also to develop new sources of energy, greater supplies, and to reach out to the underdeveloped world. We should not take resources from the "have-nots" for our own selfish needs, so to speak, without recognizing that the gap between the "haves" and the "have-nots" in this world is widening rather than narrowing. This is the most explosive single element, in my opinion, threatening the peace of the world.

In short, I feel that we who represent such a small part of the world population have a moral responsibility not to consume energy at rates that take it away from other parts of the world.

SENATOR RANDOLPH: Well, Mark, comment on Libya. What has it done in recent days?

SENATOR HATFIELD: You go ahead; you and I are of one mind on this.

SENATOR RANDOLPH: Well, yes, that is right. Libya is taking over the oil company properties, and country after country is, I think, on the threshold of doing the same.

SENATOR HATFIELD: Which we can look at, politically, as a threat to our security. But again I would like to emphasize the point: I don't think we have a right to take, take, take, for our own enjoyment, for our own purposes, without recognizing the impact of this on the rest of the world.

SENATOR RANDOLPH: We, of course, give dollars to buy their oil. But the heads of the governments don't always use that money for the benefit of their people.

SENATOR HATFIELD: That is right.

LETITIA DAVIS, Environmental Affairs, Department of Health, Education and Welfare: We've heard a lot tonight about higher energy costs. Won't this be another example of a relatively heavier burden falling on our country's poor? I am wondering if you see any way of counteracting this regressive effect. You have mentioned the possibility of changing the rate structure on electricity consumption. Do you see any other changes?

SENATOR RANDOLPH: I share with you, Letitia, a strong commitment to do what we can for those who have lesser incomes. I feel that very strongly. Today the President signed a bill to rehabilitate the handicapped in this country. That's just another measure for persons who need help.

However, I think we have to realize that taxes in the United States are lower than those in any other industrial country of the world. Now that doesn't take care of the situation you describe. But, when comparisons are made, American taxes are really not at the top.

SENATOR HATFIELD: This is not only true in the field of energy. Any time we begin to talk about constricting consumption, we are always going to find that the impact falls heaviest upon the poor.

That was one subject taken up at the conference on environmental questions that met recently in Stockholm. Representatives of the developing countries argued that it was very easy for the industrial countries to talk about environmental policies to restrict economic growth and development because they had already reached the stage that the developing countries were hoping to reach. In other words, the developing countries were not ready to adopt the restrictions that we in the industrial nations were suggesting.

I think Senator Randolph put his finger on the best way at the problem: we have to be creative in these other areas—education, rehabilitation, and so forth—but most especially it comes back to that necessary balance between economic needs and environmental quality. Let's not forget that some of the loudest promoters of environmental restrictions are members of the affluent class whose hobby it is to create and support ideas like this. Now these people *are* contributors and I am not demeaning them at all. But to some extent their attitudes reflect the fact that they are also comfortable.

Let me give you a quick example. We once had a witness before our committee who told us not only that we should *not* build any more dams on the Middle Snake River, but also that we should yank out all the dams we built. Senator Bible, who was presiding, turned to me and asked, "What did he say?" I replied, "I think he said 'yank out the dams'." Senator Bible said, "You mean including Hoover Dam, all of the dams?"

I asked the witness if he had ever seen floods and people who had been inundated and suffered. "Oh, yes," he said, "I have flown over such places in my airplane." And I said, "Well, sir, surely you use electricity. I am sure you realize rural—." He said, "Oh, I have my own private generators." Well, the whole picture was that of a king in the castle on the hill looking down upon all of the peasants in the flood plain who had to fend for themselves.

Then he made this astounding statement: "I am convinced the world was better off when we had famines, floods, and pestilence." There sat a man who enjoyed all the great benefits of our society and he was saying we were better off when we had floods, famines, and pestilence.

That attitude is much more prevalent than a lot of us are willing to recognize.

WILLIAM C. ADAMS, Standard Oil Company of Indiana: Senator Randolph, you and Senator Hatfield have both stressed the importance of protecting the environment while at the same time serving the country's energy needs. I would like to ask Senator Randolph for his thoughts on the respective roles of federal and state government in formulating such a balanced policy.

SENATOR RANDOLPH: Of course, industry, business and commerce must be included in whatever is done. Government programs at any level can't do the job all by themselves. We must turn to the people that have the know-how and that have been in these activities over the years and work with them. Rather than downgrading the contributions of industry and business, which some people are tempted to do, we must bring these elements together and work together.

I am an environmentalist and a conservationist, but above all else, I am a realist. We no longer can march down two roads, the environmental road and the energy road. If we are to do the job, we have to bring the two together. Also, I am not a defeatist. I don't run to the wailing wall and I don't press the panic button. I'm convinced the American people are going to have the intelligence to tackle this problem by making use of all of the available components that are necessary to solve it— and certainly industry is one of these.

EDWARD L. BEACH, Senate Republican Policy Committee: My question is for Dr. Ray. We have been talking about alternatives, and it seems to me that we should at least consider alternative sources of energy that haven't been adequately exploited yet and maybe are free for the taking if we just knew how to do it. For instance, has any work been done on trying to make use of the tides, of the winds of the earth—which can do tremendous damage and therefore should be able to do a lot of good, too—of changes in barometric pressure and of the ocean currents.

SENATOR RANDOLPH: Windmills.

DR. RAY: First of all, let me say there is no such thing as a free lunch. Any source of energy will cost something. Just because the sun shines up there doesn't mean that we can collect this energy and convert it to space heating or electricity, or whatever, without costs. There will be costs, and in some cases they will be considerable.

For example, calculations show that, given present technology, if we were to make the arrays of solar cells needed to collect enough of the sun's energy for a central power station to produce electricity, say, at a thousand megawatts, the costs would be three to four times what they are for any of the fuel sources we presently use.

DR. McCRACKEN: That is, equipment costs would be three to four times higher than they are for today's conventional generating plant?

DR. RAY: That is right. And, of course, the amount of land area that would be occupied would be enormous.

On the other hand, this technology is already feasible for some things, for example, heating and perhaps cooling homes and one-story buildings. The main thing lacking is the manufacturing capacity to produce the components. This hasn't developed because the market hasn't been there, but it may begin to develop now.

With respect to the other energy alternatives, of course windmills have been used for certain purposes for a long time. There have been some interesting new technological developments in the design of windmills, mak-

75

ing them much more efficient aerodynamically, and now work is under way to develop enough windmill power to run a small generating plant.

It is difficult to use winds, as well as tides and ocean waves, for energy because they are intermittent, they do not come at the time when you have peak demand and, except for tides, they are not dependable or predictable. So, in each of these cases, we get back to the problem that we do not yet have a good way of storing energy.

Now, there is a tidal power plant in operation off the coast of Brittany in France. The tides there are enormous, of course, second only to those in the Bay of Fundy. The plant does generate electricity, but it is not economic.

Also, there has been a plan for some time to develop a tidal power station in Maine, taking advantage of the enormous tides there. The difficulty that has held it back is not so much the requirement for developing reservoirs to hold the water that comes in with the tide and for letting it spill down through the generators, as it is to find a way, because of the intermittent nature of the force, of storing the power. So here again we come back to the need for an R & D program. We hope that there will be more work done in developing better storage capacity, both in terms of batteries and other ways of storing energy, so that if it can't be stored as electricity, it can easily be converted to something else and back again. In other words we need storage techniques that will make possible a continuous flow of power even though the source of power generation is intermittent.

One more comment about the tides. The movement of the tides back and forth across the shores of the earth causes a certain amount of friction against the land underneath. It is possible to do a little calculation and show that the force involved here, the energy involved, is equal to all of the energy utilized at the present time in the entire world. But, you see, the force is very diffused, and what you need in order to generate electricity with it is to concentrate it. That is the hard thing to do.

DR. McCRACKEN: By the way, on this matter of dependability, I recall some hot days in July back on the farm when I was a boy when the windmill was not very dependable and I had to pump the water.

MR. ADAMS: Let me suggest to Dr. Ray the approach of a big enough grid with three main points, or locations, where the tides offset each other. Tides are predictable, and one could choose the locations so that one tide was going up, one was going down, and one was in between. Then you could keep pumping the juice in from different locations.

DR. RAY: Yes—but you'd have the problem of long distance transmission, which is another area where we have great losses. That too would require more R & D.

Let me say one thing further on the predictability of tides: yes, indeed they are, but for one year at a time. Did you ever stop to think that we never predict tides more than one year ahead?

SENATOR RANDOLPH: Dr. Ray, I thought the poet said, "One ship drives east and another drives west by the self-same winds that blow." What about it?

DR. RAY: The winds are not predictable, but you can, by the application of good aerodynamics and skillful seamanship, sail into the wind.

SENATOR RANDOLPH: The set of the sail?

DR. McCRACKEN: Let's end this Round Table on that poetic note.

We appreciate very much having our panel with us this evening—Dr. Philip Trezise from the Brookings Institution, Senator Jennings Randolph of West Virginia, Dr. Dixy Lee Ray of the Atomic Energy Commission, and Senator Mark Hatfield of Oregon—for this extremely lively discussion of what is one of the most important and difficult problems of public policy today. [Applause.]

PART III

DOMESTIC AND INTERNATIONAL ISSUES

Paul W. McCracken, *Moderator*

❧

J. William Fulbright
John N. Nassikas
George W. Ball
Charles J. DiBona

PAUL W. McCRACKEN, Edmund Ezra Day university professor of business administration, University of Michigan, and Round Table moderator: This country faces a very difficult energy problem, one that has come upon us so fast that most of us are still a little bewildered. It is quite clear that the problem is posing very complicated difficulties for us, both at home and in our foreign policy. Tonight's task is to explore the interface between our domestic energy problem and our external political and economic relationships.

Senator Fulbright, would you make a few comments on that subject?

J. WILLIAM FULBRIGHT, United States Senate (Democrat, Arkansas): The effects upon our external relations, both economic and political, are very far-reaching, I think.

I am sure Mr. Ball will deal with economic aspects, but let me just point out that we are threatened with the prospect of an enormous deficit in our balance of payments because of the necessity of purchasing oil from abroad.

The energy problem also, of course, is focusing attention on the situation in the Middle East and on our relations with the countries of that region where most of the oil is. About 75 percent of the world's oil, roughly, is in the Persian Gulf and in the North African countries.

There are some good and bad aspects to what's happening in this area. One bad aspect is that the enormous wealth being developed in certain countries is causing some of the leaders to become more militant than they have been before. On the other hand, I think the situation is strengthening the position of a man like King Faisal of Saudi Arabia who has traditionally been a friend of the United States and has been most cooperative, in my view, under the circumstances. It has also, I think, resulted in his being much closer to the Sudan and Egypt.

There has been, in addition, a noticeable change recently in the attitude of Mr. Dayan in Israel. I believe the Israeli government is recognizing the significance of oil to the United States. Within the last few days, Mr. Dayan has made a proposal—it was reported in the press —indicating a softening of Israel's attitude toward the negotiation of some kind of settlement of the Arab-Israeli conflict.

Overall, there are many other effects too. One I should mention that I don't like at all is the enormous flow of arms into the Persian Gulf. This is said to create stability. But I wonder if it isn't more likely to create the means by which much more violence will come about in the course of border incidents and the other things that could occur there. I have often thought about the recent minor incursion into Kuwait from Iraq. If Kuwait had possessed all the arms it is now purchasing and had had them ready to go, then very likely, instead of a very mild reaction and finally negotiations, the government would have responded with missiles and tanks and there would have been quite a little war going on there.

But, anyway, that decision has been made and we are pouring arms into the Persian Gulf. I think it could be dangerous.

Another aspect of the situation is that it is drawing attention to the vulnerability and dependence of our country and others—particularly, of course, Japan which is far more dependent than we are on Middle East oil. And I put the oil problem together with some other recent developments, for example, the shortage in food. Just this morning there was a very ominous news article about what is happening in the fishing industry all over the world, and particularly off our own coast.

Now, as to energy, which is in many ways even more

basic than food, I think if we haven't lost our capacity to reason altogether, we will recognize these relationships and the interdependence of countries with respect to energy. That should cause us to look with much greater favor upon the United Nations, for example, and to understand the necessity for negotiating settlements of these issues. Nothing could be more disastrous to our economy and to our whole industrial enterprise in this country than a war breaking out, say, in the Persian Gulf. A serious interruption of production there, even though we are not as dependent as Japan and Europe, would be a major disaster today, far greater than the first Suez closing back, I believe, in 1956.

So a good side of the situation is that it is impressing upon people the absolute necessity of finding ways to negotiate our differences. Our political differences are so sensitive that it has been almost impossible to make progress. But now with the fundamentals of food and energy involved, I think that surely if we haven't taken complete leave of our senses there will be movement toward much greater cooperation.

Well, there is much more that could be said, but I think that is enough to get us started.

DR. McCRACKEN: Mr. Ball, you have observed this problem and participated in discussions and negotiations about it, both as under secretary of state and then as an investment banker. You must have some views on this.

GEORGE W. BALL, senior partner, Lehman Brothers: Well, I have views, but I also have a good deal of confusion. The confusion results from the fact that there are many different sets of statistics and forecasts because there are many variables involved in this total situation.

Senator Fulbright mentioned the projected impact on the United States' balance of payments of our increasing dependence on imported oil, particularly on oil from the Middle East. To estimate that effect, one has to make assumptions about the price structure, about the degree to which alternative sources of energy may be developed and the timing, and so on.

83

If one accepts some of the figures that have been put forward, the U.S. import bill for oil may be something like $18 billion by the early 1980s. This could be a very alarming figure, if it is taken as representing a net burden on the United States' balance of payments. But the fact is that, by and large, after one nets these figures by subtracting items such as transport expenses and so on and then makes some assumptions about what the Arab states may do with their surplus revenues, my guess is that the figure will be quite manageable. Instead of $18 billion, it may be in the neighborhood of $6 to $8 billion, or something of that kind—which, given the improvement of the United States' trade position as a result of the change in currency parities, we ought to be able to live with.

We also face the problem of the excess earnings of the Arab states. The fact is that certain of the Arab states, particularly Saudi Arabia and Kuwait, are going to have—on almost any projection—earnings very far in excess of their absorptive capacities. The question is—what will they do with vast sums which are above and beyond the amounts they can spend?

They may be tempted to leave the oil in the ground for what may be quite well-considered conservation reasons. If they do, it will create problems of scarcity for the rest of the world. On the other hand, if they maintain production, there will be pressures to put their excess earnings into short-term investments—which could add to the instability of the monetary system. A great deal of short-term speculative money floating around could create problems. Hopefully, a great part of the money can be invested, or at least part of it can be invested, in improving the level of life of the Arab peoples—not only in the producing states, but in the other Arab states, so that the whole Middle East can begin to blossom like a rose as some areas of it already have.

But I think it is a very big question. We may very well see a flow of capital from the Arab states into Western Europe and the United States in the form, not merely of portfolio investment, but of direct investment.

In my judgment, this would probably be a healthy thing.

But beyond the purely economic aspects of this, I am frankly somewhat more concerned than Senator Fulbright —who took what, to me, seems to be a reasoned but also a very optimistic view of what might happen. I am not at all sure that if powerful countries start scrambling for scarce energy resources, they will be sensible enough to join together. Past experience indicates that situations of scarcity tend to bring out the worst instincts of people and nations rather than their better instincts. And there are very great elements of division in the whole energy situation.

Let's assume that some of the Arab states—the major producing nations—build up resources of money far beyond their capacities to spend in any rational way. To some extent, in my view, this would make the more conservative Arab states much less able to resist the arguments of the more activist Arab states that they should use their oil for political purposes. If, for example, Saudi Arabia builds up huge excess reserves, then, given the dynamics of the Arab world, the pressure from the radical Arab states to use oil as a political weapon in the conflict between Israel and the rest of the Arab world would be, I think, very strong indeed.

It is my guess that there will be some yielding to that pressure, to that temptation, and that inevitably oil will be used as a political weapon—though not necessarily explicitly. The explanation for withholding oil is likely to be a mixture of the legitimate considerations of conservation—that oil should be kept in the ground if it is producing revenues for a country in excess of its capacity to spend—and the strong suggestion that if the Western world, and particularly the United States, were to change its attitude toward Israel, the Arab states would feel a greater obligation to meet the soaring requirements of the Western world for the oil that comes out of their soil.

If this should happen, and if there should be a slowing down in the growth rate of oil production or indeed a cutback by some of the major producing states—given the fact that production and requirements are just about even today—I would expect to see great elements of division. If the action were discriminatory against the United States, Americans would resent the fact that Western Europe was getting its full quota when the United States was not. If the action were even-handed, I think you would find the countries of Western Europe in short supply and, with a much heavier dependence on oil to run their industries than we have, they would bring great pressure on the United States to adopt a new policy toward the Middle East.

If the President of the United States ever has to explain to the American people that they have to endure rationing of heating oil and gasoline for automobiles because the Arabs refuse to accept the American policy toward Israel, the result would be terribly divisive and nasty for the domestic scene. It could produce very ugly reactions.

So I foresee a good deal of difficulty arising from all of this. Although I hope that Senator Fulbright is right, that there can be agreement amongst the oil-consuming states to avoid a nasty fight in which each tries to undercut the other to ensure its own supplies of energy, I'm not sure this will be the case. I think it is going to require a very high order of statesmanship, on the part not only of the United States, but of the other consuming nations as well.

DR. McCRACKEN: Mr. DiBona, since you are in the White House you must look at this from more than the dispassionate view of a bystander. How does it look from your point of view?

CHARLES J. DiBONA, special consultant to the President, Executive Office of the President: Well, I think that one has to put the question of fuel imports in perspective and look at the problem over time. At present the United States imports about 15 percent of its energy. This is radically different from Japan which imports well over 90 percent of its total energy supply.

DR. McCRACKEN: That is, we import 15 percent of our total energy, not just of oil?

MR. DiBONA: Our total energy—and a little over a third of our total oil.

Of the oil that we import, under 10 percent comes from the Middle Eastern countries, so that at the present time our dependence upon the Middle East is not large.

What troubles people is the future. According to projections, oil imports are expected to grow very, very rapidly, so that by the early 1980s we will be importing over 50 percent of our oil, a great portion of it from the Middle East. And that is cause for concern.

The question is, what do we do about it? It seems to me we can do two things—reduce energy consumption in the United States and increase our domestic production. The administration has taken a number of steps in that direction, and we think they will have real effect.

As Mr. Ball mentioned, it is difficult to predict precisely where we will be by the 1980s, but this country's basic energy resources are vast:

- We have half of the total world supply of coal, which is a very cheap energy resource. Our coal supply is enough to last us at least three centuries at present consumption rates, so we could radically increase its use. There are difficult environmental problems of course, but we can do something about that.
- We have enormous quantities of oil locked in the shale of Colorado, Wyoming and Utah, as much oil as there is in the Mideast.
- We have developed nuclear-powered plants and are installing them at a very rapid rate.

There are a number of things we can do. The President is, for example, advocating a ten-year program of $10 billion for energy R & D which will, we believe, significantly reduce our dependence upon foreign sources by the 1980s. Moreover, we are embarking now, in part because of the problems we expect this winter, on an aggressive program of energy conservation. I think that certainly by the time we get to the 1980s the price of energy resources in the United States will have risen to a point where energy utilization will be a good deal more measured than it is or has been in the past.

So, I think there is some light at the end of the tunnel. The problem is going to be in the next several years when none of these new measures will, because of the long lead times involved, have brought about significant increases in domestic supply. So, for the next several years we will be increasingly dependent upon traditional sources of energy. Then hopefully, as a consequence of these other measures, this will taper off.

MR. BALL: May I say, Dr. McCracken, that I agree that there is light at the end of the tunnel. I agree also that there is no long-term energy shortage in the United States, because we are sitting on vast amounts of potential energy. The question is not whether there is light at the end of the tunnel, but what happens in the middle of the tunnel, and this is what I was really' commenting on a few minutes ago. I think the problem is from now until, say, the middle or latter 1980s; this is the period when these political pressures and developments could take place in a way that could be extremely disturbing and disruptive.

DR. McCRACKEN: John, I dare say that, as chairman of the Federal Power Commission, you must be sitting there in the middle of the tunnel, aren't you?

JOHN N. NASSIKAS, chairman, Federal Power Commission: Sometimes.

DR. McCRACKEN: How does it look from your vantage point?

MR. NASSIKAS: Sometimes it looks as though there is no light at either end of the tunnel.

At the same time, I think that in the short term—let's take from now until 1980—we can do a great deal through governmental policy to improve our energy situation and to reduce somewhat our reliance on imports. It is perfectly evident, however, that we will have to import both petrolum products and crude oil over the short term, the next seven years—and an increasing share of it from the Middle East. But for that short term, as Mr. DiBona pointed out, we do have abundant coal resources. The major portion of those resources is recoverable only by deep-mining. However, some coal deposits are near the surface and can be strip-mined. About two-thirds of the strippable coal and lignite are located west of the Mississippi.

Insofar as we can, we should convert some of our oil-fired plants to coal-fired plants. If we convert, however, the conversion has to be based upon a non-appealable final decision that the coal can be burned and utilized if it meets primary air quality standards. There has been no assurance, at least to date, of such a non-appealable final decision.

DR. McCRACKEN: You mean otherwise a company just would not feel justified in making the investment to finance the switchover?

MR. NASSIKAS: Yes, I don't believe it would be justified in doing so in the absence of that kind of a guarantee.

Also, in the near term we should get our nuclear power program back on schedule. We are behind about 30,000 megawatts—that's 30 million kilowatts. Well, what does that mean? It's about as much power as is developed in New York, Pennsylvania, New Jersey, Maryland, Delaware and Washington, D.C. This is the amount of power that is not onstream from nuclear plants which were scheduled to be in operation two years ago. Each nuclear plant of a thousand megawatts that we can get onstream will displace an enormous amount of fossil fuel for electric power generation. Today the electric utility industry consumes about one-fourth of all of our primary energy consumed in the United States.

DR. McCRACKEN: By primary, you mean coal, oil, gas?

MR. NASSIKAS: Coal, oil and gas, primarily. By the year 2000, if we use modest growth forecasts and if electric power will share in the growth, probably 50 percent of our primary energy consumption will be by the electric utility industry. So, I cannot overemphasize the necessity in both the near term and the long term of getting our nuclear program back on schedule.

DR. McCRACKEN: What has been the trouble?

MR. NASSIKAS: It's complex. Part of it has been environmental, part of it has been a regulatory process that can't cope with the adversary system for determining whether a plant should be certificated, part of it has been quality control by manufacturers, and part of it the question of safety.

I believe, nevertheless, that the major problems can be overcome so that nuclear plants can be placed onstream within a reasonable period of time—and not the decade that it takes today from the time a plant is planned until it is onstream. In Japan, the identical plant can be generating electric power in five years, about half of our time.

As to coal, I cannot overemphasize that we must not have legislation that forbids strip-mining. We should be able to strip-mine coal, provided that the land can be reclaimed and provided that the environment can be preserved.

And I am in full agreement with you, Charley, that much can be done on conservation. I know that much more can be done to conserve gas and electric power, which are within the FPC's area of responsibility, than we have done in the past. Just one example, we are burning enormous quantities of natural gas under boilers for electric power generation. The conversion efficiency is about 32 percent, compared to about 50 or 60 percent efficiency on a direct application for space heating. This should cease.

The Federal Power Commission has established priorities and also economic constraints to reduce the burning of gas as boiler fuel in the event other fuels are available.

DR. McCRACKEN: But how are you going to do that? The heavy demand for gas by industrial users has come about, I assume, because of the price structure we have. How are you going to change that?

MR. NASSIKAS: Yes, part of it is directly attributable to a price structure which makes gas more than competitive with all other fuels. Natural gas costs about 32 to 33 cents per million BTUs currently for boiler fuel use for the generation of electric power, compared to 38 cents on average for coal and up to 60 to 70 cents per million BTUs for low sulfur oil.

The FPC is changing some of the rate design to the extent that it has jurisdiction. I would like to remind the audience, Mr. Chairman, that the FPC does not have jurisdiction over intrastate use. About 70 to 80 percent of our intrastate gas consumption is for industrial use and electric generation, with about 60 to 70 percent of the total going for boiler fuel.

MR. DiBONA: I might add that one of the most important pieces of legislation that the administration has sent to Congress this year, I believe, is the proposed Natural Gas Supply Act. This would deregulate and free the price of new gas supplies made available to interstate pipelines. It is tremendously important because this fuel, which is really the cleanest fuel we have, a very premium fuel, is now very much underpriced because of present regulatory arrangements.

The result has been tremendous demand and very little development of new reserves. This has brought about large curtailments in the delivery of the gas—which will add to heating oil problems if it should be cold this winter.

There is probably no more significant fact, I think, than the curtailment in natural gas deliveries as a consequence of regulation which keeps the price so low that there is little incentive to go out and drill for new supplies. We have copious evidence to indicate that there are very large quantities of natural gas in the United States. It will cost more to develop it, but we could bring in these supplies. Currently, the average price of natural gas in the United States is about 20 to 30 cents a million BTUs. At the same time, because of the shortage of gas, we are importing gas at close to a dollar a million BTUs—but our regulatory arrangements don't permit a U.S. producer to charge even half that.

It is a ridiculous arrangement, one that has to be changed. The matter is in the hands of the Congress, and is probably the most important short- and medium-term step that can be taken to ease our present problem.

SENATOR FULBRIGHT: You say it is in the hands of Congress. Congress passed a bill for deregulation of natural gas in 1956 but the President vetoed it. So don't lay responsibility for this on the Congress. [Laughter.]

MR. NASSIKAS: Well, this President is taking a different approach. Mr. Chairman, I just want to make it crystal clear that we are constrained by a regulatory statute as to the price level we can establish for natural gas. The rates have to be based on costs, not based on commodity values. The Natural Gas Act is a consumer protection statute, and it has resulted in giving consumers the lowest price for the highest premium fuel in the United States. Regrettably, however, consumers are running out of service because prices are so low that not enough new reserves are being developed. We need a statutory change in order to permit efficient allocation of this resource through the price mechanism.

However, price alone will not allocate the resource. We also have to accelerate lease sales even beyond the acceleration which the administration has recommended. On the present schedule, lease sales won't reach 3 million acres annually until about the year 1978 or 1979. Instead we have the entire Atlantic outer continental shelf and the Gulf of Alaska under study and the study won't be completed until next June. In the meantime, not a single exploratory well or developmental well, no core drilling, will occur on the entire Atlantic coast.

According to the U.S. Geological Survey, about 65

trillion cubic feet of gas are off the Atlantic coast. The states that are suffering most from the gas shortage are the eastern states, which are also the states most strongly opposed to this kind of program. I favor it, and I think we ought to get on with it.

DR. McCRACKEN: Would it be fair to say that although obviously no one likes to pay higher prices for gas or anything else, the consumer's stake in this is nothing less than whether the gas is going to be available to him for his house?

MR. NASSIKAS: That is true—because the imported gas is not a displacement of domestic gas, but rather a supplement to it, and we need all of these sources. So, when we import gas at a dollar per million BTUs, that just adds to the gas supply that will be used for industrial use. We set up incremental pricing on these imports—I know you would be interested in that, as an economist—so that those who want to pay a dollar had better have the market for it.

DR. McCRACKEN: I think we should award you a Ph.D. in economics.

MR. NASSIKAS: Thank you kindly.

DR. McCRACKEN: Obviously, one of the things that people are very much interested in is the question, what can a typical consumer or typical family expect this winter and next summer in terms of such practical things as the availability of heating oil and gasoline? What is the ordinary family facing in the relatively near term, the next twelve months?

MR. DiBONA: The Department of Interior has just completed a rather detailed study of the winter heating oil situation. The figures show that if we have a normal winter—that is, if it is neither colder nor warmer than normal—and if the refineries operate at their current high levels of output and if there are no problems with the de-

93

livery of crude from abroad as well as from American wells, then we will need to import about 650,000 barrels of heating oil per day in order to meet normal needs.

To put that figure in context, I might note that the most the United States has ever imported for an extended period is 530,000 barrels per day, and that was for the first three months of last year when all of the import controls were taken off. So, the estimate for this winter is a little over 100,000 barrels a day more than the most that we have ever imported before. It is going to be very difficult for us to do that. If it is warmer than usual, we will be in a little better shape. If it is colder than usual, that number could go up to 800,000 barrels per day.

DR. McCRACKEN: What does the weather bureau tell you about this?

MR. DiBONA: Well, we have checked with everyone, but we sort of have to rely on the *Farmer's Almanac*, and it says it is going to be cold. [Laughter.]

I might add that we are very concerned about this import situation and are taking a number of steps. The first has been to ensure that the pricing policies of the Cost of Living Council are not an impediment, a disincentive, to the import of heating oil into the United States.

The second step, we believe, must be relaxation of the sulfur standards for the burning of fuel in the United States. This action will have to come principally from the major oil-burning states, which are generally in the eastern seaboard and the north of the country. We have invited a number of governors in and spoken to them about this problem, and the President has personally spoken to them about it. If we get substantial variances from current sulfur standards, we think that we can add up to about 300,000 barrels a day to imports—through two devices:

One, Europe has a substantial quantity of heating oil that has considerably higher sulfur content than is presently permissible in the United States, so we could import additional oil from that source. Two, in order to meet the environmental standards for the heavier oils that are burned in certain areas under certain applications, such as electric generating plants, Caribbean refineries mix the light heating oils (low-sulfur content) with these heavier oils (high-sulfur content). We believe that we could get 200,000 barrels a day of distillate or heating oil out of that source if we could import and burn the higher sulfur oil.

So, it is a clear trade-off this winter, if it is cold, between the quality of the air and the warmth of people's homes.

DR. McCRACKEN: Is there going to be rationing?

MR. DiBONA: There are no plans at the moment to impose consumer rationing. However, we of course have contingency plans in preparation.

MR. NASSIKAS: Dr. McCracken, the Federal Power Commission has, in effect, been rationing natural gas through curtailment plans for upward of two years. Our prediction for this winter is that we are in an emergency, one of crisis proportions. As of last week, seven or eight days ago, we issued a policy statement and an amendment to it which permits companies that are in curtailment—that is, those who cannot meet the demands of their customers for gas—to either exchange or buy gas on an emergency basis without receiving a certificate of public convenience and necessity from the Federal Power Commission, as is prescribed under Section 7 of the Natural Gas Act.

This new policy is now in litigation. It was instituted because of the imminent emergency. We believe that this policy will not only attract gas from the intrastate market, but will also induce sales of gas that might not otherwise be made.

Overall, on a national basis, we project a curtailment equalling about 7 percent of demand this coming winter—that is, demand that won't be satisfied. In round figures this amounts to about 0.7 trillion cubic feet of gas this winter, compared to about 0.45 trillion last winter.

So, the increase is about 50 to 60 percent in our curtailment—

DR. McCRACKEN: That is, the gap is that much larger?

MR. NASSIKAS: Yes, the gap is that much larger, which means simply that larger boiler fuel users that have alternate fuel-burning capacity, like some electric power plants, must burn an alternate fuel. If they don't have an alternate fuel, they obviously will have to continue burning gas, if they can get it.

But the curtailment is there on the lowest priority uses—and of course the highest priority uses are residential and human-needs customers. There will be no homes that will be without natural gas this coming winter —none, in the entire United States. However, new customers who want to get a firm contract for natural gas may have a very difficult time in securing a supply.

DR. McCRACKEN: That is a very important statement.

MR. DiBONA: Let me point out an interesting relationship. In terms of heating value, that 0.7 trillion cubic feet of natural gas not available this winter converts to just about the level of imports that we have to make of heating oil. So, you can see the clear relationship between the natural gas situation and the heating oil situation, and the critical and crucial importance of deregulating new supplies of natural gas in order to bring in this gas.

MR. NASSIKAS: That is true, and incidentally, because of environmental standards, the quantities of available oil which meet sulfur content restrictions have been increased slightly by blending low-sulfur heating oil with higher sulfur oil in the proportions necessary to meet air quality standards.

DR. McCRACKEN: I would like to turn our attention back, if I may, to what we were talking about at the beginning of our discussion. Senator Fulbright and Mr. Ball both made some suggestions about the implications of this problem for the foreign policy of the United States. It seems quite clear that even with fairly rigorous measures here on the domestic scene, we are still going to have a pretty tight situation. So we should do everything we can to ensure reasonable security of supply. What does this mean for foreign policy?

SENATOR FULBRIGHT: The number one problem, I think, is to settle the war in the Middle East so that we can make reliable contracts, reliable agreements, for the oil and ensure political stability in the area. The war settlement, as you know, has been very tentative.

Just in passing I note that there seems to be some movement by the Israeli government. I draw this conclusion from the recent statement of Moshe Dayan that he would consider a change in Israel's position on the Sinai. At least it indicates some lessening of the rigidity, some willingness to talk about positions other than those the Israelis have insisted upon all this time.

A Middle East settlement is number one. The idea that has been booted about by some columnists and others—that in the last analysis we can always take the oil by force—is very dangerous indeed. Too often people on our side bring up what we did in Iran when Mossadegh was there, or suggest the use of force. This only adds to the fire, to the danger, and does not promote peace at all.

With regard to Mr. Ball's statement, I want to say that I qualified my so-called optimism. I said, "assuming we haven't lost our minds altogether." [Laughter.]

And I am not sure that we haven't. I can't stand up and give any assurance that we have any sense at all as to how we manage our economy or our public affairs. But I am assuming we will rise to the occasion when we have to. I haven't concluded that we are incapable of good sense, but we haven't exercised it now—as I think Mr. Nassikas has demonstrated.

The real problem in the gas business was brought about by legislation, not by the industry. I think you agree with that. It was the federal government's legislation in controlling the price, making it so utterly and unrealistically low as compared to other fuels that it drove people to use gas when they had no business using it.

97

MR. NASSIKAS: Well, legislation, plus the way it was regulated in the 1960s also—at a very low price.

SENATOR FULBRIGHT: Yes, but the basic authority for doing it was in the legislation.

MR. NASSIKAS: The basic power is there, and then the power that we—

SENATOR FULBRIGHT: Congress gave you the power, and we shouldn't have.

MR. NASSIKAS: Over the years efforts have been made to enact legislation to correct the problem.

DR. McCRACKEN: And the legislation was vetoed.

SENATOR FULBRIGHT: That is right, it was vetoed. I sponsored the bill in the Senate and Judge Oren Harris, former congressman and chairman of the House Commerce Committee sponsored it in the House. It was passed, but then it got mixed up with a scandal, if you remember. Senator Case, not the present Clifford Case but rather Francis Case from South Dakota, said he had been offered a contribution which he regarded as unethical. This caused President Eisenhower to react and to veto the bill, a most unfortunate and regrettable action.

Well, back to foreign policy. There are other areas, of course, of lesser importance in which we can use diplomacy that shouldn't be beyond our capacity. But the Arab-Israeli issue is the key problem. It is mixed up in our domestic politics; it is mixed up in our East-West relations; and just yesterday a House committee, as you know, took a vote which, if it stands, could undermine the President's whole program in seeking a détente with Russia. That agreement that was negotiated could well collapse if the action of the committee vote stands. This isn't unrelated to the problem in the Middle East, in the sense that if East-West détente fails, I don't know why we shouldn't expect a revival of Soviet interest, for example, in Egypt.

About two years ago the U.S.S.R. withdrew its troops from Egypt. We had complained bitterly about those troops and they withdrew them. I think the Egyptians expect us to be more forceful. But if East-West détente breaks down, then I think we will see an increased Soviet pressure in the Middle East, and there will be more trouble.

So, my seeming optimism assumes, as I said, some use of reason in our overall international relations—which are not looking too good at the moment, to a great extent due to the Congress. I don't think you can lay this to the executive branch.

DR. McCRACKEN: Well, I think we ought to let you have equal time here, Mr. Ball.

MR. BALL: I only said that I was extremely happy to see Senator Fulbright in an optimistic mood. I wasn't suggesting that he was wrong or right, and I hope he is right.

I do have the feeling, however, that the problems we face in the Middle East, although to an extent a test of American diplomacy, are also far beyond the reaches of American diplomacy. There are so many factors here, so many different interests involved, that the United States can only do a certain amount.

I hope I am not right that the new situation of the oil-producing countries, in which they will have more money than they can absorb, will probably result in their being pressed, perhaps even against their will, into using oil as a weapon of diplomatic policy. But if I am right, I am not sure there is a great deal the United States will be able to do about the problem, other than to try to ensure that the divisive effects aren't too widely felt.

SENATOR FULBRIGHT: May I ask you, in that connection, why did we not respond about a year ago to an inquiry by the Saudis about downstream investment and special arrangements? I never understood why we were so negative on that request.

MR. BALL: You are referring to the statement made by Sheik Yamani, the oil minister of Saudi Arabia.

SENATOR FULBRIGHT: Yes—about a year ago, not the one last spring.

MR. BALL: Well, as I recall, he suggested that the Saudi Arabians be given a special position as far as oil imports were concerned.

SENATOR FULBRIGHT: Correct.

MR. BALL: Which would have distinguished them from the other Arab producing states.

SENATOR FULBRIGHT: He also suggested an opportunity for investment in this country.

MR. BALL: Yes—that Saudi Arabia would be prepared to make downstream investments in the United States and that this would ensure it a stake in the United States, which would be an element in making for continuity of oil supplies. I thought at the time, as did many people, that this was a very constructive proposal on the part of Saudi Arabia. I can understand the position of the United States government, however, in not being prepared to give the Saudis a special position.

SENATOR FULBRIGHT: Well, why not? We have a very special position in Saudi Arabia. ARAMCO is entirely owned by American companies.

MR. BALL: Yes, but if we give a special position to Saudi Arabia, what do we do for Iran, for example, which is a major supplier of oil and which is not involved in the Arab-Israeli conflict to anything like the same extent as are the Arab states? Or what do we do for Kuwait, or Libya, and the others? It seems to me that this did present a serious problem.

Now, as far as downstream investments are concerned, I may say I personally think this approach is a very good

100

one. I speak a little bit from special interest because I played a small role in arranging one of the first major downstream investments, the one that Iran has done with the Ashland Oil Company.

Looking toward the future, I think that it will be very useful if we can persuade the Arab states to put excess funds in direct investment in downstream oil facilities, both refineries and distribution in the United States—the more, the better. I like to see the eggs scrambled this way, because then each country has a stake in the prosperity of the other—which is a desirable situation. But, to the fullest extent possible, we must avoid doing it on a basis of discrimination because that would just complicate our relations with the states of the Middle East.

What I fear, however, is that many of these issues are rather far beyond any initiative by the United States. We can't pursue these policies on our own. We can't, for example, ensure that there won't be a scramble for scarce oil resources among the various consuming nations. We can only take some lead in trying to induce a cooperative spirit so we don't cut one another's throats.

SENATOR FULBRIGHT: The scramble is already on.

MR. BALL: It is already on, but it can become very much more intense, particularly if some of the Arab states start to slow down the rate of growth in their production, because we desperately need the increased production which they are capable of.

Now, at the same time, I think that we have seen some signs of greater understanding of the problem on the part of the Arab states. It seems to me that Mr. Sadat has tried to lead the UAR into a more tractable position and that King Faisal desperately wants a stable and peaceful situation in the Middle East. They are under terrible pressures, both of these men. I also think that the Israeli government has, as you suggested, shown some signs of a greater recognition of this problem, but its position seems to me to be rather ambivalent because the Israeli Socialist party is strongly pressing for the development of the occupied lands—which can't help but exacerbate the difficult situation.

I think that the government in Jerusalem must make an effort to use its best efforts to change its present posture —which appears to the world, rightly or wrongly, and I am not passing judgment on it, to be excessively untractable. It often appears as if the Israelis have pursued procedural policies that have prevented any kind of negotiation from taking place.

Now, there has been fault on both sides, and I am not suggesting that this is either an Arab or Israeli problem. The problem on the Israeli side is that the people in the government are extremely brilliant, but they don't agree among themselves. It is their inability to formulate an initial negotiating position which makes it difficult for them to move toward a negotiating posture. Nevertheless, they must, because Israel is going to find itself in a terribly uncomfortable and difficult—indeed, even dangerous—position unless some kind of negotiating process gets under way.

SENATOR FULBRIGHT: I think it is in their interest to do it.

MR. BALL: And I think we should encourage it. I say this not as one who is at all unsympathetic with the Israeli predicament, but as a friend of Israel. I think Israel really must do this in order to give its friends something that they can defend—and up to this point that has not been forthcoming.

SENATOR FULBRIGHT: I agree with that.

DR. McCRACKEN: Thank you very much, gentlemen. We shall now take questions from our distinguished guests.

JOHN OSBORNE, *The New Republic*: Two questions for Mr. DiBona and Mr. Nassikas.

One, what actually is the profit position and the capital position of the companies involved in producing and distributing natural gas? Are they actually so desperate that they really require deregulation at the wellhead? And number two, with reference to the possibilities of reducing consumption of critical fuels, is this a real possibility in the future? In short, how much is government prepared to do, and what practically can it do, to reduce consumption in the critical areas?

DR. McCRACKEN: John, do you want to comment on that first?

MR. NASSIKAS: I will comment if I may on the profit question.

First, certainly the companies which the FPC regulates in the petroleum and gas industry are not suffering from lack of profit. In fact, if they are a natural gas company as defined under the Natural Gas Act, we must, as a matter of law, grant them a reasonable return on their capital investment. So that by definition, if we are doing our job, the companies are making a profit in addition to meeting their requirements.

The big issue, however, is not a question of raising prices so that companies will make greater profits, but rather raising prices in order that there will be greater investment in exploration and development—particularly as leases on the outer continental shelf are granted. If we continue with the present structure, costs have gone up and so prices will go up, whether we deregulate or not.

If there is deregulation, certainly there has to be a way—I will be the first to say this—to monitor the situation so that higher prices will not yield windfall profits to the producers that are involved.

MR. DiBONA: May I add a point or two on that? There are about 3,700 producers of natural gas in the United States. The industry is not highly concentrated, not dominated by a few large companies. Rather it's very diversified and spread out among many, many small producers.

Any producer, before he undertakes the drilling of a well which may or may not produce gas, calculates the expected return. What decontrolling the price of new supplies of gas would do would make it economically attractive and profitable to drill structures in areas which are not now attractive at the present price. Therefore, we would get increased supplies of gas. We have rather extensive studies which show that the market for natural gas would clear—that is, demand and supply would be balanced—at around 50 cents per million BTUs (or per thousand cubic feet). So, that is the economic incentive that is available and that we should be directing principally at the very small producer who is, in fact, the wildcatter. He is the man who goes out and takes the chance on new structures. It is worth noting that generally the way gas wells, and oil wells too, get into the hands of large companies is that they are bought after the discovery is made.

With regard to your second question, that of conservation, we are not sure exactly how much fuel we can save. Our estimates suggest that we could manage about 100,000 barrels per day in reduced consumption of heating oil this winter through an aggressive public campaign.

DR. McCRACKEN: What would that be percentage-wise?

MR. DiBONA: About 3 percent of the heating oil used in the United States.

This summer, we witnessed an interesting phenomenon. The increase in gasoline consumption for the first five months of this year, 1973, was running as predicted, about 7 percent ahead of the corresponding month of the previous year. Then in June that dropped to about 2 percent, and it has been running at about 1½ to 2 percent above last year for each of the succeeding months.

104

So, clearly, there has been a reduction in gasoline consumption as a consequence of public concern about the problem, and we think that there is some real hope here.

ARNOLD PARKER, Committee for Economic Development: My question is for Senator Fulbright and Mr. Ball. Let's assume that U.S. dependence on Mideast oil is reduced substantially, but Europe's and Japan's remains, and that prices for Mideast crude go up. To what extent would a reduction in American dependence not matched by reductions elsewhere lessen our security problems?

DR. McCRACKEN: Senator Fulbright?

SENATOR FULBRIGHT: I am not sure I understand the question. I will let Mr. Ball try his hand on this.

MR. BALL: I would say, first, that a decrease in American dependence on Middle Eastern oil would mean a decrease in the total requirements of the consuming nations for Middle Eastern oil—which would mean, in turn, that if the producing states slowed their rates of expansion, the resulting shortage wouldn't be as abrupt or as immediate or as serious as would otherwise be the case. In other words, it would mitigate the problem, not eliminate it.

I would think that we ought to do everything possible to reduce our dependence on Middle Eastern oil, recognizing that we will not succeed in the short term, that we are going to have problems, and that those problems are almost certainly going to take a political turn.

SENATOR FULBRIGHT: Well, for a country just to decrease its dependence through these enormously costly crash programs, some of which have been suggested recently, doesn't make much sense to me. The other approach of trying to seek a diplomatic solution to the problem and utilizing Middle East oil, pending the development of these more long-term alternatives that Mr. Nassikas talked about, makes a lot more sense to me. The suggestion that we should become self-sufficient at great cost, at a much greater cost than using the oil that is there, doesn't seem to me to be a very sound policy.

MR. BALL: I wouldn't at all suggest that we try to become self-sufficient because, in the first place, we couldn't manage it for very many years to come. But I do think that there are some things we ought to think about seriously.

For one, we ought to try to understand the position of the producing states in the Middle East. A problem each of them faces is that no matter how large the reservoir of oil that lies under its soil, that reservoir is still finite. Even a nation like Saudi Arabia, which is more blessed with oil resources than any other nation in the world, still has to look to the time in the far distant future when its reserves begin to decrease.

Now, what should these countries do about it? At present they have no other significant means of earning foreign exchange, no other means of sustaining themselves. To some extent American technical assistance, the infusion of American capital, the guidance that we can supply in helping them use their own money to develop other sources of revenue—to industrialize, to become great fertilizer producers, to use their oil resources to produce other kinds of commodities, to use what manpower resources they have—these things can help. Of course, Saudi Arabia doesn't have much, but some of the other nations do.

All of this can be very beneficial. Take the situation in Iran. Iran isn't at all likely, even if it wished to, to take a political position that would curtail its supply of oil to the world, because it is engaged in development programs which use all the financial resources it can generate. Iran is very different from Saudi Arabia. It has a population of 30 million compared to less than 6 million for Saudi Arabia, so that the situations are not comparable.

Nevertheless, I would say that a very great effort ought to be made by the United States to help these nations develop alternative sources of revenue. To the extent that this is achieved, they will feel less concerned about pumping the oil out of the ground so fast that they have to face a period of depletion.

SENATOR FULBRIGHT: What you are saying now is much more sensible to me than to go all out on a highly expensive crash program.

MR. BALL: But I would say this, Senator Fulbright: I think we have to do both. I am not sure that crash program is quite the right description, but we are certainly going to have to move toward more effective utilization of the energy resources that we have so that we—

SENATOR FULBRIGHT: I agree with that. In the long term, you are quite right. The use of all these alternatives should come into the picture.

There seems to be a feeling that is evidenced in certain statements of my colleague, Senator Jackson, that we have to have self-sufficiency right away, that we just cannot trust these irresponsible people. I thoroughly disapprove of that kind of a policy. Instead we ought to move in the direction you are describing.

MR. DiBONA: I might speak a little bit more about this problem of pricing. We do not have the option of getting low-cost foreign oil or high-cost U.S. production of oil and coal and other resources. The facts of the matter are that today it is significantly more expensive to import a barrel of oil into the United States than it is to produce a barrel of oil right here—by over a dollar per barrel. The price of oil has been going up sharply. We are in a difficult bargaining position and are facing very rapid increases in price.

The cost on the spot market of the marginal or extra barrel of oil now being imported from the Persian Gulf, the cost of procurement plus the cost of transport, is about seven dollars a barrel. At seven dollars a barrel it is very competitive to produce oil from shale or to liquefy coal.

Our choices are going to be high priced energy abroad and high priced energy domestically.

DR. McCRACKEN: In other words, what you are saying at this point is that the prices confronting customers down the way a bit are just going to be higher?

MR. DiBONA: Yes, sir, and I would add another point: It is important for us to develop our resources in a way that maximizes possibilities for reducing the cost of these conversions. In the future, probably the only thing that will put a lid on the price of foreign oil will be the availability of large supplies of alternative energy, and that is what we have to provide. But it doesn't mean we should stick stubbornly with the alternatives. If producers will sell oil cheaper, we will take it cheaper.

SENATOR FULBRIGHT: I agree with that, except that we can't do this with shale oil at the moment.

MR. DiBONA: Well, I would agree that—

SENATOR FULBRIGHT: And we don't have any plans to do it.

MR. NASSIKAS: I agree with you about shale oil. But if we get on with the leasing program, we can, between the time that we sell the leases and start to move gas through the pipelines—well, even with environmental impact statements and all, we still should be able to get some productivity within three years.

SENATOR FULBRIGHT: Oh, I am entirely for that.

MR. NASSIKAS: That is short term, you see; that is short term.
Now another point that has not been made tonight and that should be made, because it is important, is that every barrel of oil that we import means less gas in the United States. Why? The reason is that gas and oil are a product of joint exploration.

About one-fourth to one-third of all the gas in the United States has been discovered in the course of a search for oil. Taking the latest statistics we have, which are right up to date, the exploration and development of gas wells in the United States is up about 30 percent compared to a year ago, whereas exploration for oil is down by about 6 percent. So if we import more oil, the trend of less gas as a by-product of the search for oil is going to continue.

I cannot overemphasize the point that in order to reverse the energy situation in the United States, we must be certain that we have a massive exploration and development program for petroleum, in addition to gas. Otherwise, we won't get the gas we need.

SENATOR FULBRIGHT: Well, I certainly don't want to delay what you are talking about. I was thinking of technological things like how to convert shale oil economically and how to treat coal so that it meets pollution standards. That takes some time, as you know.

MR. NASSIKAS: I agree with you.

SENATOR FULBRIGHT: You have put your finger on the difficulty. It is the short term that we are worried about, more than the long term.

WILLIAM BURGESS, Ohio State University: Could someone discuss what, if anything, is being done to use economic incentives, pricing systems and other equivalent means for reinforcing and channeling decentralized self-administering decisions in order to bring about lower energy consumption, particularly in the home and places like that?

MR. NASSIKAS: We have a number of programs at the Federal Power Commission which are designed to reduce demand and to allocate resources more effectively.

In the field of electric power, the FPC has jurisdiction over about 15 percent of the kilowatt hours in the United States. The remainder is under the state jurisdiction; but we do have national responsibilities through a uniform system of accounts by which we can exert some leadership with the state commissions—and we try to do so.

We have attempted to establish rate designs which will inhibit the wasteful use of natural gas. We have attempted to cycle the utilization of electricity, insofar as we can, over a twenty-four hour period, or through diversity exchanges. We have a vastly improved interconnected electric power system in the United States which enables diversity exchanges to be made among approximately thirty-two power pools in the United States.

In the field of natural gas, as I mentioned earlier, we have been trying to deter the use of gas in, say, electric power generation. And we have succeeded in doing this because about four months ago oil passed gas as a power generation fuel for the first time. Coal is still the dominant fossil fuel here, accounting for about 40 percent of our power generation by fossil fuel. We also have the gas curtailment programs that I noted earlier. And we have a rate design under the so-called "Seaboard formula" whereby we are attempting to allocate natural gas more in terms of amounts consumed (the commodity side) and less in terms of the fixed charges associated with the capacity to provide service (the demand side). All I mean by that is that the more an industry uses, the more it will pay if the rate design is tilted toward the commodity side rather than the demand side. This is a very promising approach.

So the FPC is using economic policies in an attempt to allocate energy resources more effectively.

DR. McCRACKEN: We have exhausted our time here tonight—if not our huge subject. May I express appreciation to the members of the panel—Mr. Charles DiBona of the White House, Senator J. William Fulbright, Chairman John Nassikas of the Federal Power Commission, and Mr. George Ball, former under secretary of state.

This concludes the last of AEI's three Round Tables on the energy crisis. [Applause.]